19.50

500 PLANTS
OF SOUTH FLORIDA

500 PLANTS

OF

SOUTH FLORIDA

By JULIA F. MORTON

FAIRCHILD TROPICAL GARDEN

Miami, Florida

Julia F. Morton, D. Sc., F.L.S., is Research Professor of Biology, and Director of the Morton Collectanea of the University of Miami. All photographs in this book were taken by the author.

The line drawings on the end papers are reductions of the full-page illustrations by Dr. Frank D. Venning which appeared in 400 Plants of South Florida.

For
John DuPuis
whose special delight
is in his plants

Foreword

DURING THE PAST 20 years, Nature and Man have made many changes in the native and exotic plant population of South Florida. Some pioneer plant collections have been sadly depleted by hurricane and cold damage; some have been wholly replaced by realty developments, as have extensive stands of natural vegetation. Many splendid and precious specimens have given way to progress and one too often experiences a pang of regret when passing a site where a favorite tree once flourished.

On the other hand, there is compensation in the great enhancement of our horticultural resources that has taken place in this period. Nursery and landscaping services have multiplied and plantsmen have been exceedingly active in introducing or developing plants for home and commercial beautification. Many species which were rare or unknown in our gardens in 1952 are now commonly seen in dooryards and along streets and highways.

Having such an enriched inventory, we should by now have reached a stage of maturity in our ways of utilizing the materials available. Unfortunately, too much planting is still done on a trial-and-error basis; many still indulge in fast-growing plants for quick effects and are sorry later when their temporarily gratifying plants loom too big too soon and require major operations of reduction or removal. Valuable time is thus lost during which desirably compact and slow-growing plants, properly situated, could have been gradually developing, becoming more beautiful and satisfying each year, with a minimum of maintenance. Our ornamental plants should give us a maximum of pleasure, not problems. With our ever-diminishing living and recreational space, we should be extremely careful to choose plants we can happily live with, evaluating each not only by its present appearance but by its likely future performance and the effort that may be required to maintain it as an asset to our property and to the community.

In preparing this revision of 400 PLANTS OF SOUTH FLORIDA,* it has been necessary to add 150 species, drop 50 of the less important from the first edition, and alter and amplify nearly all of the 350 retained. Furthermore, there have been numerous revisions of nomenclature, placing affected plants in a different alphabetical position in the text. In a few cases, erroneous names formerly in general use (for example, *Cydista aequinoctialis* for the GARLIC VINE) have been abandoned and the currrently approved botanical name substituted. Some plants have been shifted to different families. As not all botanists are in accord, I have sought to avoid controversy in this matter by following the 7th edition of J. C. Willis' *Dictionary of the Flowering Plants and Ferns* (1966).

For sentimental reasons, I am inclined to preserve whatever of the original edition need not be changed. The FOREWORD, especially, I prefer to reproduce intact in memory of my co-author, Dr. R. Bruce Ledin, and my late husband, Kendal Morton, whose collaboration and companionship were lost in 1959 and 1964, respectively.

The style and purpose of the book remain the same—to provide the layman with the essential facts on as many as possible of the more popular or interesting plants of South Florida, without making the book too lengthy or cumbersome for use and enjoyment.

Coral Gables, Florida.
January 1974   JULIA F. MORTON

*by J. F. Morton and R. B. Ledin; Text House (Florida) Inc., 1952.

[8]

Foreword to the First Edition

"400 PLANTS OF SOUTH FLORIDA"

THE TREES, shrubs, vines and herbaceous plants encountered in subtropical and tropical South Florida are today so numerous and varied that the queries, "What is the name of this plant?", "Where is it native?", "What is its value?", are frequently voiced, not only by newcomers but also by long-time residents. As an aid in answering these questions, brief, non-technical descriptions of more than four hundred plants of South Florida have been prepared and assembled in this book. These include the majority of the exotic and native plants cultivated in home gardens and parkways, as well as the more noteworthy of the wild plants seen by the wayside, on the beaches, and on the Florida Keys.

Florida is endowed with a greater number of native species of trees than any other state in the Union. In addition, South Florida, particularly, has received a wealth of plant introductions from many countries whose climates, in whole or in part, are tropical or subtropical. The influx of plants to this area, apart from the transport of seeds by birds, ocean currents, and storms, doubtless began with Indian migrations. The Spaniards brought, among other plants, the Sour or Seville Orange, and early settlers from the West Indies, especially from the Bahamas and Cuba, brought many of their favorite fruits and ornamentals. Later, as the state became more populous, rare ornamental and economic plants were imported by nurserymen and others. Greatest in number and variety, however, were the importations which came through the activities of the famed plant explorers of the United States Department of Agriculture, many of whose early discoveries in faraway areas were sent to the Plant Introduction Gardens in Brooksville and in Miami, and were later distributed to experimenters and nurseries for trial. And still the plants come, through both public and private effort, for interest in plant introduction seems contagious where the climate affords so much opportunity for the enjoyment of outdoor life.

While not a specific geographical section of the state, South Florida is commonly understood to include an irregularly-shaped portion of warmer winter climate, the northern border of which extends from Merritt Island, on the east coast, southward and westward around the lower end of Lake Okeechobee, then northward to Bradenton on the west coast, with isolated northern extensions in water-protected St. Petersburg and Clearwater. All of this area is generally accepted as subtropical, despite occasional winter frosts, and the claim is frequently made that the lower east coast and tip of the peninsula are actually tropical. On the Florida Keys, the winter temperature is further moderated and, in southernmost Key West, the climate is undeniably tropical and the plant life of special interest. This latter area, unique in the United States, was made the subject of a detailed survey, and virtually all of the plants found there are included in this volume.

No attempt has been made to cover the plants of northern Florida, which are typically hardier species, though many of the plants dealt with will be found there and even further north. Neither has it been considered desirable to include species largely limited to the grounds of experiment stations and other collections of rare specimens, lest the book grow to unwieldy proportions and thus be less useful to the visitor, gardener, and plant lover for whom it is intended. Also excluded are annual flowers which do not form a part of the established vegetation but come and go with the seasons. Grasses, other than bamboos, and most of the weeds are also omitted.

A number of publications which deal with the native and exotic plants of Florida, and which are in print or generally available in libraries, are listed at the end of the text. These are suggested reading for those who wish to learn more about Florida's plants.

The authors are deeply indebted to Dr. Frank D. Venning, In Charge, Swingle Research Project, University of Miami, for his pen-and-ink drawings and they wish to acknowledge with thanks the help of Mr. Roy Woodbury, Assistant Professor of Botany at the University, in identifying specimens. Thanks are also due Mr. Kendal Morton, Director of the Morton Collectanea of the University, and husband of the senior author, for his encouragement and his constructive reading of the manuscript.

Coral Gables, Florida.   JULIA F. MORTON
January 4, 1952.   R. BRUCE LEDIN

500 Plants of South Florida

THE PLANT DESCRIPTIONS prepared for this volume are deliberately simplified "word-pictures." As stated in the foreword, they are meant to aid the layman in recognition or confirmation. In presenting details of form and structure, technical terms, required for botanically exact expression, have been carefully avoided. The sizes given are those that may ultimately be attained by the species in their native areas, since the sizes of plants are variable factors and the specimens in South Florida are of various ages and changing statures.

The botanical, or scientific, names (consisting of genus and species, and sometimes the variety) provide the only means of stable identification, though even these may occasionally be revised, and one given preference over another. The plant descriptions which follow are accordingly arranged alphabetically under their currently preferred botanical names. Some have a number of synonyms; only those in general use or employed by present-day authorities are given and included in the index. For readers not familiar with the abbreviations following the botanical names, it should be explained that they represent, in most instances, the name of the botanist who conferred the name on the plant in question. Even the botanical name, followed by a different author-abbreviation, will often apply to another species. The appendage "Hort." is not an author-abbreviation but indicates that the species has been accepted by horticulturists though its origin may be unknown. The abbreviation "Auth." indicates that the name is commonly used by many authors.

At the right, opposite the botanical name of each species, appears the name of the plant family to which it belongs. The worker with plants wishes to know that certain plants are related. This knowledge is also used in seeking further information in books in which plants are grouped under their family names.

The common names shown are those which seem most popular and which are most frequently encountered in writings. No attempt has been made to include all common names, for they vary with different localities. Indeed, the reader is cautioned not to be influenced by common names alone, for that which is applied to a plant in one area may be applied to a quite different plant in another. All cited are found in the index, which will direct the reader to the descriptions against which the plants' characteristics may be checked.

Edibility, medicinal, poisonous and other characteristics and uses are mentioned in many cases. The uses are largely primitive ones prevailing where the plants are native or commonly grown. Some have become widely adopted, but not necessarily in Florida.

To aid the gardener in plant selection and culture, the rate of growth, usual means of propagation, light requirements, ornamental value or, in some cases, undesirable features, are briefly stated.

Sweet Acacia—*Acacia farnesiana*

500 Plants of South Florida

Acacia auriculaeformis A. Cunn.   LEGUMINOSAE
EARLEAF ACACIA—Native to Australia. Tree, to 50', upright, not spreading, with compact head of slender, curved, leaf-like blades 5 to 7" in length; flowers bright-yellow, very small, in 2 to 3" clustered spikes; fruit a pod up to 4" long, coiled, containing flat, black seeds attached to pod by orange, string-like arils. Tree fast-growing from seed; salt- and drought-tolerant; ornamental, and having beautifully-grained wood.

Acacia farnesiana Willd.   LEGUMINOSAE
SWEET ACACIA; POPINAC; OPOPANAX; CASSIE—Considered native to tropical America but widespread in the tropics and grows wild in South Florida and on the Keys. Shrub or a small tree, to 12', bushy and thorny; foliage feathery; flowers tiny, golden-yellow in round clusters ½" across, very fragrant; fruit a brown, cylindrical pod up to 3" long. Seeds sprout readily. Perfume is made from the flowers, a black dye or ink from the fruits; the unripe fruits are used medicinally and a gum similar to gum arabic is secured from the stems and branches.

Acalypha hispida Burm. f.   EUPHORBIACEAE
CHENILLE PLANT; RED-HOT CATTAIL—Native to the East Indies. Shrub, to 15'; leaves up to 8" long, broadly-oval, with pointed tip, tooth-edged; flowers in hanging, furry "cattails" up to 1½' long, deep-red and borne profusely and continuously. Foliage edible when young. Grown from cuttings, in full sun or light shade.

Acalypha wilkesiana Muell.-Arg.   EUPHORBIACEAE
COPPER-LEAF; MATCH-ME-IF-YOU-CAN; JACOB'S COAT—Native to the South Sea Islands. Shrub, to 15'; leaves up to 8" or more in length, soft, heart-shaped with pointed tip, tooth-edged; color of leaves variegated, may be green-and-red, maroon-and-pink, entirely red or rich bronze. Flowers tiny, red, in short spikes. Foliage edible when young. Cuttings of mature wood may be placed directly in the ground. Most colorful foliage in full sun. Defoliated by severe cold snaps.

Acoelorrhaphe wrightii H. Wendl. (*Paurotis wrightii* Britton)   PALMAE
EVERGLADES PALM—Native from southern Florida to the Bahamas, Cuba, Central America and southern Mexico. Palm, to 40', forming large clumps; leaves to 3½' wide, fan-shaped, deeply divided into slender segments, pale or silvery beneath, on spiny stalks; flowers yellowish in erect panicles to 3½' long; fruit nearly round, black, ⅜" wide, single-seeded. Slow-growing from seed but many have been successfully transplanted from natural swampy habitat to high ground, a practice now restricted by law. Salt-tolerant and highly effective in landscaping where there is sufficient space for this elegant native palm.

Acrocomia armentalis Bailey (*A. crispa* C. F. Baker)   PALMAE
COROJO—A Cuban palm 35 to 45' tall, with bottle-shaped trunk covered with long, dark spines until old, when it is nearly smooth. Leaves feather-like, to 9' long, with fine, needle-like spines on the rachis and leaflets. Flower spathe, to 5 ft. long, has very spiny stalk. Fruits round, to 1 ¼ " wide, with edible pulp. Seeds yield a medicinal oil. Leaf fiber made into fly whisks and cordage.

Adansonia digitata Linn.   BOMBACACEAE
BAOBAB—Native to tropical Africa. Tree, to 60' with enormously swollen trunk, to 60 or even 100' in circumference, and open, spreading branches; leaves deciduous, compound, the 5 leaflets spread like fingers, lance-shaped, to 5" long, downy beneath; flowers (summer) to 6" wide, suspended on 1 ½'-long stalks; the 5 sepals and 5 white, waxy petals curled back, exposing the thick, white, staminal tube terminating with a dense ball of stamens tipped with golden anthers; fruit oblong or oval, 2 to 6" thick and to 1' long; woody, covered with gray-green felt and filled with dry, white, acid pulp and hard, kidney-shaped seeds. The acid pulp is made into a cooling drink, the leaves are dried and pulverized for soup; in fact, the weird tree and all of its parts fill many needs, especially of the desert people. Grows readily from seed and lives to a great age, perhaps thousands of years, in Africa. There are a number of fruiting trees scattered around South Florida.

Adenanthera pavonina Linn.   LEGUMINOSAE
RED SANDALWOOD TREE—Native to tropical Asia. Tree, usually up to 25', may become very large; bark light-gray; leaves semi-evergreen, pinnate, leaflets up to 2" long; flowers small, yellow or white, in elongated clusters 6" or more in length; fruit a brown pod to 9" long, coils after opening and contains red, flat-oval seeds called "Circassian seeds", used in necklaces and other novelty-wear also as weights for measuring precious metals; edible when peeled and roasted, somewhat intoxicating raw; wood used as a substitute for sandalwood. Grows at moderate rate from seed or cuttings. Seed slow to germinate because of hard coat.

Adenium obesum Balf.   APOCYNACEAE
DESERT ROSE—Native to East Africa. Shrub, succulent, 6 to 10' high, with smooth, thick trunk, stubby branches and gummy, white, poisonous sap; leaves deciduous, clustered at branch tips, fleshy, oval, glossy, to 5" long; flowers in terminal bunches, funnelform, 5-lobed, 2 to 3" wide, deep-pink or red-and-white; seed pods in pairs, to 9" long, containing seeds attached to silky hairs. Slow-growing, drought-resistant. Cultivar *multiflorum* fairly recent in Florida nurseries.

Baobab
Adansonia digitata

Agapanthus africanus Hoffmgg.   ALLIACEAE
AFRICAN LILY—Native to South Africa. A vigorous, attractive, bulb-plant, forming clumps to 2' high. Leaves leathery, erect, to 10" long and ½" wide. Flower stalk bears 12 to 30 blooms, usually light-blue (may be white or dark-blue) which hold up well as cut flowers.

Agave neglecta Small   AGAVACEAE
WILD CENTURY PLANT; BLUE CENTURY PLANT—Native to South Florida. Succulent plant, formed by a rosette of strong leaves 3-4' long and 6" wide, rising from base and curving outward, bluish-gray, spiny; flowers on erect central stalk 30 to 40' tall, each bloom about 2" long, yellowish or greenish, ill-smelling. Blooms once when several years old and then dies. Propagated by offsets.

Agave sisalana Perr.   AGAVACEAE
SISAL AGAVE; SISAL HEMP; GREEN AGAVE—Native to southern Mexico. Succulent plant, with rosette of green or grayish, sword-like leaves rising to 6' from base, usually smooth except for spine at tip; blooms once and dies; flower stalk ascends to 30' and bears on short branches numerous flowers, each 2½" across, greenish and ill-smelling, which produce young plants that take root when the stalk falls. Introduced into Florida by Dr. Henry Perrine in 1836 and later cultivated in many tropical countries for its fiber. Watery juice irritates skin.

Sisal Agave—*Agave sisalana*

Lebbek Tree—*Albizia lebbeck*

Candlenut
Aleurites moluccana

Yellow Allamanda
Allamanda cathartica

Copper-Leaf
Acalypha wilkesiana

African Lily
Agapanthus africanus

Violet Allamanda
Allamanda violacea

Woolly Morning-Glory
Argyreia nervosa

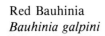

Red Bauhinia
Bauhinia galpini

Flame-of-the-Forest
Butea frondosa

Aglaonema modestum Schott   ARACEAE
CHINESE EVERGREEN—An herb, from Kwantung, China, with multiple slender
stems ("canes") rising up, 1 ½-2', from a spreading rhizome. Leaves deep-green, oval,
pointed, to 1 ½' long and 5" wide, are leathery, waxy and arched or drooping.
Inflorescence inconspicuous. A popular foliage plant, easy to maintain in pots indoors. *A. commutatum* Schott, probably from the Moluccas, has leaves streaked
across with light green. These plants are propagated by cuttings and air-layers; should
be cut back when they become leggy.

Albizia lebbeck Benth.   LEGUMINOSAE
LEBBEK TREE; WOMAN'S TONGUE—Native to tropical Asia and northern
Australia. Tree, to 100', with wide-spreading branches and "nuisance" roots which
travel far out from the tree; bark pale; leaves deciduous, pinnate, with leaflets to
1 ½" long; flowers (spring) small but in fluffy greenish-yellow clusters, so abundant
as to produce a showy effect, followed by long, flat, dry, buff-colored pods which
rattle in the breeze. Wood dark-brown, good for furniture, sometimes called East
Indian Walnut; seeds, flowers and bark used medicinally; bark used for cleansing.
Tree drought-tolerant, susceptible to wind-damage; fast-growing from seed; springs
up like a weed in neglected lots.

Aleurites moluccana Willd.   EUPHORBIACEAE
CANDLENUT—Native to Malaysia and common in most tropical regions. Tree, to
60' or more, "snowy" from a distance due to silvery-white powder on new foliage and
twigs which eventually turns "rusty". Leaves evergreen, variable; may be wedge-
shaped or 2- to 5-lobed; range from 3 to 15" across. Flowers small, white, in large
sprays. Fruit grayish-brown, 2-2 ½" wide, with 1 or 2 irregularly ridged seeds which
are hard-shelled, black with whitish waxy coating. Kernel white, somewhat toxic raw,
not too digestible cooked. Kernels may be strung and burned for illumination, also
yield oil for lamps or for making candles, soap, etc. Whole nuts, rubbed smooth,
buried in marshes to season, then polished, serve as handsome, jet-like, beads for
necklaces and bracelets. Fast-growing from seed.

Allamanda cathartica Linn.   APOCYNACEAE
YELLOW ALLAMANDA—Native to Brazil. Shrub, climbing; leaves evergreen,
elliptic, to 6" long, in groups of 3 or 4, glossy; flowers (spring to fall) bell-shaped, 5-
lobed, brilliant-yellow, white-throated. Variety *williamsii* Raffill has somewhat hairy
leaves and flowers to 3" wide with brownish throat; variety *hendersonii* Raffill has
thick, glossy leaves and flowers up to 5" wide; variety *schottii* Raffill has slightly downy leaf-stems, large flowers, with dark-yellow throat. The bur-like fruit (to 2" wide) is
rarely seen. Grown from cuttings. These allamandas are among the most popular ornamentals in South Florida. Leaves and bark are cathartic; latex purgative in quantity, may irritate very sensitive skin.

Allamanda violacea Gardn. & Field   APOCYNACEAE
VIOLET ALLAMANDA; PURPLE ALLAMANDA—Native to Brazil. Shrub,
climbing, green-stemmed with reddish tinge on sunny side; leaves evergreen, slender,

pointed, to 6" long, in whorls of 3 or 4, light-green, rough on top, pale beneath with hairs on midrib and veins; buds glossy, maroon-purple at tip; flowers (summer) bell-like, 5-lobed, 3" long and up to 2½" wide, dull-magenta with faded areas giving a 2-toned effect, few in a cluster at tips of branches or in a series of clusters to 1' long. Grown from cuttings in full sun.

Alocasia macrorrhiza Schott ARACEAE
GIANT ELEPHANT'S EAR—Native to southern Asia and widely grown in the Tropics. Herb with thick stem, to 5 or even 15' high. Leaves erect, long-stemmed, the "arrowhead" blade 3 to 8' long and up to 4' wide, glossy, entirely green or irregularly blotched with white. Contains irritant, acrid juice. Stems and starchy corm toxic raw; have served as emergency food after repeated and prolonged cooking. Propagated by division.

Aloe barbadensis Mill. (*A. vera* Linn.) LILIACEAE
BARBADOS ALOE; MEDITERRANEAN ALOE—Native to Mediterranean region. Herb, succulent, with a rosette of upright, thick, fleshy, rubbery leaves up to 2' long and 4½" wide at base, tapering to a long point, spiny-edged; develops a stout stem when old; flower stalk (winter) rises from center to a height of 4' or more; flowers yellow, tubular, 1" long, hanging in an upright spike sometimes 1½' long, nectariferous and attractive to humming-birds. Yellow latex in green skin of leaves stains clothing brownish-red; dried, it is the laxative "bitter aloes" of the drug trade. The translucent, jelly-like pulp of the leaves is used to soothe burns, heal wounds, etc., and is steeped in water and taken internally; also is an ingredient in cosmetic products. Multiplies by offsets which are easy to transplant.

Barbados Aloe—*Aloe barbadensis*

Alpinia purpurata K. Schum.   ZINGIBERACEAE
RED GINGER—Native to the Pacific Islands. Herb, perennial, with erect, arching stems 6 to 15' high, forming clumps; leaves evergreen, alternate, elliptic, pointed to 1' long; flowers white, tubular, 1" long, each enclosed in a purple-red bract, arranged in showy terminal spikes 5 to 10" long. Propagated by division or by the plantlets which form in the old spikes.

Alpinia zerumbet Burtt & Smith (*A. speciosa* K. Schum.)   ZINGIBERACEAE
SHELL-FLOWER; SHELL-GINGER; PEARLS-OF-THE-ORIENT—Native to eastern Asia. Herbaceous plant growing in clumps of many upright stalks, to 12' high; leaves lance-shaped, up to 2' long and 5" across, glossy; flowers (spring-summer) numerous in drooping, elongated clusters, the pearly buds opening a few at a time; expanded flower about 2" long, like open mouth of snake, outer petals white with pearly luster, pink-tipped; inner petal bright-yellow, flushed and speckled with red and red-streaked on protruding lip. Plant yields fiber for cordage and paper. Propagated by division.

Alternanthera dentata Stuchl.   AMARANTHACEAE
RUBY-LEAVED EVERLASTING—Native from Peru and Brazil to the West Indies. Herb, perennial; much-branched; to 3' high; leaves oval to oblong, pointed, 1½ to 4" long; green in the wild but in the cultivar *rubiginosa* are wine-red above and purple below; flowers tiny, white, in globose heads. Grown from cuttings. Popular as a ground cover and for borders and planting strips. Has been nicknamed "Florida pussy-willow".

Amaryllis vittata Ait. (*Hippeastrum vittatum* Herb.)   AMARYLLIDACEAE
BARBADOS LILY; COMMON AMARYLLIS—Native to Peru. Herb, with onion-like bulb and upright, strap-like, light-green, smooth leaves which may or may not be present at blooming time (spring); flower stalk fleshy, cylindrical, erect, up to 2', carrying at the top a cluster of handsome, bell-like flowers to 6" long, and typically red-and-white striped. Many named cultivars of different colors are available. Bulbs may remain in ground for years or may be taken up after blooming and reset in winter, in full sun.

Shell-Flower
Alpinia zerumbet

[19]

Pineapple—*Ananas comosus*

Soursop—*Annona muricata*

Ananas comosus Merr. (*A. sativus* Schult.)　　　　　　　　　BROMELIACEAE
PINEAPPLE—Native to tropical America. Herb, growing in rosette form with
several tiers of tough, slender, sharp-pointed leaves curving outward and downward from the central stalk; may reach 4' in height; flowers purple, small
and densely arranged on a central cone which develops into the pineapple; fruit compound with rough, "tiled" rind, yellow, reddish-orange or dark-green when ripe;
flesh white to yellow, very juicy, sweet to subacid. Fruit topped by a crown of tiered
leaves. Plant usually has barb-edged leaves (except Smooth Cayenne variety); fruits
when 1 to 2 years old, the main shoot bearing only once. New plants grown from
crown, from slips which develop at base of fruit, or from suckers or "ratoons" at base
of plant. Miniature pineapples now grown commercially as house plants.

Annona glabra Linn.　　　　　　　　　　　　　　　　　　ANNONACEAE
POND APPLE; ALLIGATOR APPLE; incorrectly CUSTARD APPLE
(q.v.)—Native to South Florida, the Bahamas, West Indies and tropical America;
also tropical Africa. Tree, to 45'; leaves evergreen, somewhat oval, with pointed tip,
glossy; flowers up to 1" cross, whitish, or greenish-yellow, with dark-red markings;
fruit (nearly all year) conical, to 5" long, smooth, rich-yellow skin; pulp salmon-
yellow in dryish segments, with undesirable, musky, subacid, resinous flavor; seeds
brown and bean-like. Some fruits of fair quality; most are eaten only by wild animals.
Tree prefers swamps (where the trunk becomes 3 to 4' in diameter) but can be grown
on moderately wet or high land.

Annona muricata Linn.　　　　　　　　　　　　　　　　　ANNONACEAE
SOURSOP; GUANABANA—Native to tropical America. Tree, to 20', slender and
erect; leaves (nearly evergreen), narrow, up to 4" long, dark-green, glossy and
pungently aromatic; flowers whitish, fleshy, on short stems; fruit may be 10" in length
and 6" broad, elongated-heart-shaped or irregular oval, sometimes with a curved tip;
skin green and covered with soft spines or projections; flesh white, cottony, very
juicy, subacid to acid, aromatic; seeds black, bean-like. Strained pulp excellent mixed
with milk and sugar as cold drink and makes delicious ice cream. Fruit borne on
branches or trunk, sometimes close to ground, continuously except in winter.
Propagated by seeds or grafting on own rootstock. Popular in Key West; survives
only in protected locations on mainland; defoliated, if not killed, by cold snaps.

Annona reticulata Linn.　　　　　　　　　　　　　　　　ANNONACEAE
CUSTARD APPLE; BULLOCK'S HEART; JAMAICA APPLE—Native to
tropical America. Tree, to 25', somewhat spreading; leaves deciduous, lance-shaped,
to 7" long; flowers greenish or yellowish with purple markings; fruit (spring) broadly
conical or heart-shaped, up to 6" long, yellow-brown or clear yellow, often with
bright-red or orange-red cheek; pulp creamy-white, custard-like, sometimes granular,
sweet; seeds numerous, brown, bean-like. Hardier than the soursop and rather common on the older homesteads of South Florida; fruits often mummified by insect-
attack.

[21]

Annona squamosa Linn.   ANNONACEAE
SUGAR APPLE; SWEETSOP—Native to tropical America. Tree, to 20', slender
proportions, rounded head; leaves evergreen, narrow, to 4½" long, dull; flowers
greenish-yellow on short stems, sometimes 3 or 4 in a cluster; fruit (summer) heart-
shaped or rounded, up to 4½" across; rind thick, composed of segments which
separate when fruit is ripe; yellowish-green or sometimes bluish-green with whitish
bloom; pulp in segments, creamy, sweet and delicious; seeds bean-like, dark-brown
or nearly black, shiny. Fairly common but fruit often mummified by insect attack.
Grown from seed (pre-soaked); budding and inarching of selections is practiced in India. Seeds of this and other species poisonous.

Anthurium andreanum Lind.   ARACEAE
FLAMINGO FLOWER—Native to Colombia. Herb, with long-petioled leaves rising from an underground stem and forming clumps to 3' high; leaf blade elongated
heart-shaped to 1' long and 6" wide. Flowers minute, white or red, in a narrow, cylindrical spike, with a waxy, heart-shaped, red, pink, green or white bract (to 6" long) at
its base. Inflorescence persists for 2 months on plant, 1½ months when cut for flower
arrangements. Thrives in raised beds or pots in semi-shade. Usually grown from
suckers. If the plant climbs, it should be beheaded, the top planted and cuttings made
of the stem.

Antidesma bunius Spreng.   STILAGINACEAE
BIGNAY—Native from the foot of the Himalayas to northern Australia. Tree, 30-
80', with broad, dense crown of drooping branches. Leaves evergreen, glossy, oval,
pointed, 3½ to 8" long; flowers reddish, malodorous; male, in spikes; female, in
racemes, on separate trees; fruit oval, ¼" long, single-seeded; skin smooth, turns
from green to white, then red and finally purple-black; flavor acid, blackberry-like.
Fruit clusters borne in showy profusion. Ripe and unripe fruits together are excellent
for juice, jelly and wine. Air-layered trees bear heavily in 2 years.

Antigonon leptopus Hook. & Arn.   POLYGONACEAE
CORAL VINE; CHAIN-OF-LOVE; ROSA DE MONTANA—Native to Mexico.
Vine, climbing, slender-stemmed; leaves heart-shaped, pointed, soft, light-green with
attractive veining, up to 5" long; flowers about ½" long, 5-parted; when closed,
resemble little 5-sided hearts, strung like pendants along the wavy tendrils; bright-
pink, reddish-pink or white. Blooms almost continuously. Fast-growing from tubers
or from the small seeds. Common in South Florida, escapes and covers vacant lots
and trees. Severe cold snaps turn leaves brown and unsightly.

Araucaria heterophylla Franco (*A. excelsa* R. Br.)   ARAUCARIACEAE
NORFOLK ISLAND PINE—Native to Norfolk Island, east of Australia. Tree, not
a pine though related; erect, to 200'; horizontal, frond-like branches arranged in tiers,
1 to 2' apart, up the trunk; branchlets ornamental in pattern and densely covered with
short, fine, evergreen foliage, dark-green, blue-green or white-tipped; cone round, up
to 6" across. Grown from tip cuttings. Attractive when young but fast-growing and
becomes too tall for one-story landscaping; also leans away from shade, and is blown

Norfolk Island Pine—*Araucaria heterophylla*

down by strong winds. Overplanted in past few years; some disenchanted owners have tried beheading, with unlovely results.

Archontophoenix alexandrae H. Wendl. & Drude   PALMAE
  (Ptychosperma alexandrae F. Muell.)
ALEXANDRA PALM (named for the Danish Princess Alexandra); frequently mis-named ALEXANDER'S PALM—Native to Queensland, Australia. Palm tree, to 80', somewhat swollen at base, otherwise slender and straight, ringed; leaves feather-like, arched, grayish underneath, up to 6' long with leaflets 1 ½' long and 2" wide, pointed or notched at tips; flowers whitish, in drooping, branched cluster 1' long; fruit round-oval, to ½" long, red. Seeds remain viable 2-3 months.

Ardisia escallonioides Sch. & Cham.   MYRSINACEAE
MARLBERRY; MARBLEBERRY—Native to South Florida. Shrub or tree, to 25', with slender proportions, light grayish bark; leaves evergreen, lance-shaped, up to 6" long; flowers fragrant, ¼" across in large clusters, white and purple with yellow stamens; fruit (spring) ¼" wide, round, black, glossy, one-seeded, acid, edible. Grown from seed, easily transplanted, and blooms early. Dr. Henry Nehrling tried without success to popularize it as an ornamental.

Ardisia humilis Vahl (*A. solanacea* Roxb.)   MYRSINACEAE
SHOEBUTTON ARDISIA—Native to India. Shrub or small, bushy tree, to 35', erect, compact; leaves evergreen, smooth, oblong-oval, pointed, 4-8" long; flowers ½" wide, star-shaped, mauve, in clusters among the leaves near the ends of the branches; fruits, glossy, turn from red to black, ¼" wide, oblate, like old-fashioned shoe buttons, edible. Everbearing, handsome ornamental for sun or shade; slow-growing from seed or air-layers.

Shoebutton Ardisia—*Ardisia humilis*

Queen Palm—*Arecastrum romanzoffianum*

Arecastrum romanzoffianum Becc. (*Cocos plumosa* Hook.)   PALMAE
QUEEN PALM; FEATHERY COCONUT PALM—Native to South America.
Palm tree, to 60', trunk about a foot thick, gray, smooth but ringed, somewhat shaggy
at top; feather-like leaves to 15' long, dark-green, drooping, with leaflets hanging
from midrib like a fringe; flower cluster up to 6' long; fruits orange, round, about 1''
across, sweet but fibrous. Unripe seeds poisonous; ripe seeds used for novelties.
Popular ornamental in South Florida though some homeowners object to the quantity of fruits that accumulate at the base of the trunk. Easily grown from seed less than
3 months old; thrives in full sun.

Breadfruit
Artocarpus altilis

Jackfruit
Artocarpus heterophyllus

Argyreia nervosa Bojer (*A. speciosa* Sweet)　　　　CONVOLVULACEAE
WOOLLY MORNING-GLORY—From tropical Asia and grown in all warm climates. Vine, large, woody, with twining stems reaching 45'; white, velvety down on stems and underside of leaves; leaves heart-shaped, 8 to 12" long; flowers (all summer), bell-shaped, pale-lavender-pink with dark-purple throat; base clasped by 5 greenish-white, silky-haired sepals which dry and shrink and form a scalloped saucer, whitish beneath and tan on top, holding the nearly round, brown seed pod which develops after the flower falls. Thus is formed the "baby wood rose" valued for novelties. Fast-growing from seeds (4 in each pod), or from cuttings; in full sun. Needs protection from wind.

Aristolochia elegans Mast.　　　　ARISTOLOCHIACEAE
CALICO FLOWER—Native to Brazil. Vine, with slender, twining stems; leaves broad-heart-shaped, to 4" wide; flowers, solitary, 2-3" wide, formed like a dutchman's pipe, basically ivory-white, velvety, the inside of the "bowl" spattered with reddish-purple, the throat yellow with purple border and hairy. Not odoriferous. Seed capsule brown, 6-ridged; splits open like upside-down parachute to release seeds.

Aristolochia grandiflora Sw. (*A. gigas* Lindl.)　　　　ARISTOLOCHIACEAE
PELICAN FLOWER, or DUCK FLOWER—Native from southern Mexico to Panama and the West Indies. Vine, herbaceous, climbing to 10'; leaves heart-shaped, pointed, 4-10" long; flower bud inflated, duck-shaped; open flower 6 to 20" long with pendent, ½-3' "tail"; whitish outside, mottled dark-purple and hairy within with purple throat; unpleasantly odorous when first open. Seed capsule up to 4" long and 1¾" thick. Variety *Sturtevantii* has largest flowers and is the most commonly grown.

Artocarpus altilis Fosb. (*A. incisa* Linn.; *A. communis* Forst.)　　　　MORACEAE
BREADFRUIT; INFERTILE BREADFRUIT—Native to Java and adjacent islands. Tree, ultra-tropical, to 60', with slender branches; leaves evergreen, up to 3' long, broad and deeply cut with as many as 9 lobes; male flowers tiny, closely set in a club formation up to 1' long which is cooked and eaten in some areas; female flowers in a ball; fruit round, 6" or more across, skin rough with tile-like pattern or fairly smooth, green when unripe, yellowish when ripe; pulp whitish to yellow, seedless. Cooked, unripe or ripe, and eaten as a vegetable, or made into "poi". At home only in Key West. On the mainland has survived outside for a time and there are several large specimens under glass. Propagated by root shoots or air-layers. The Seeded or Fertile breadfruit, so far as is known, has not matured in Florida. This type, having edible seeds, is incorrectly called Breadnut, a name properly limited to *Brosimum alicastrum* Swart.

Artocarpus heterophyllus Lam.　　　　MORACEAE
JACKFRUIT—Believed native to India; widely cultivated in southern Asia. Tree, 30-70' tall, with gummy, white sap; leaves evergreen, oval, to 9" long, leathery, glossy; male flowers tiny, in oblong, clublike clusters 2-4" long; female, in elliptic or rounded clusters, on same tree; fruit, compound, the largest borne by any tree, ranges

from 8" to 3' in length and 6 to 20" in width; rind thick, rubbery, studded with small, hard points, brownish-yellow and odoriferous when fruit is ripe; flesh yellow, succulent, in bulb-like sections, with pineapple-banana fragrance and flavor; seeds numerous, white, ¾-1½" long, edible when cooked. The full-grown but unripe jackfruit may be boiled as a vegetable; the ripe flesh is eaten raw, cooked or preserved. Grows slowly from seed or air-layers; needs protection from cold when young.

Sicklethorn Asparagus—*Asparagus falcatus*

Asparagus falcatus Linn.   LILIACEAE
SICKLETHORN ASPARAGUS—Native to tropical Asia and Africa. Vine, woody, spreading, to 40', with numerous short, sharp thorns; abundant leaf-like branchlets in clusters of 3 to 5, bright- to rich-green, 1½ to 3" long, very slim; flowers (winter) white, small, in sprays, profuse and fragrant; fruit brown. Grown from seed; thrives in shade.

Asparagus setaceus Jessop (*A. plumosus* Baker)   LILIACEAE
ASPARAGUS-FERN; LACE ASPARAGUS-FERN—Native to South Africa. Vine, climbing, with spiny, woody stems; very slender, stiff, green branches and minute, hair-like branchlets form graceful, sheer, fern-like "fronds"; flowers white, tiny, in small clusters; fruit red to dark-purple, round, ¼" wide. An old-time favorite as a pot plant and on porches, and the greenery is commonly used in floral arrangements. Propagated by division or seed. There are many named cultivars, some dwarf.

[28]

Asparagus sprengeri Regel (*A. densiflorus* Jessop)　　　　LILIACEAE
SPRENGER ASPARAGUS—Native to South Africa. Herb, bushy, with arching or
sprawling, spiny stems up to 6 ' long; leaf-like green branchlets in clusters of 3 to 8,
very slender, 1 ″ or more in length; flowers (early summer) pinkish, in sprays,
fragrant; fruit red, 3-lobed, ½ ″ wide. Much grown in pots or hanging baskets; also
as a ground cover in moist locations. Propagated by seed or by division.

Asystasia gangetica T. Anders. (*A. coromandeliana* Nees)　　　ACANTHACEAE
GANGES PRIMROSE—Native to Africa and southwestern Asia. Herb, with
sprawling, creeping stems up to 4' long, rather downy; leaves oval, pointed, to 4"
long; flowers (all year) 1 ½ " long, tubular, flaring to 1" wide, lavender, yellowish or
white, in clusters to 6" long. Plant cooked and eaten as greens; extract used for
medicinal purposes. Grown from cuttings or layers as a ground cover in sun or partial
shade.

Averrhoa carambola Linn.　　　　　　　　　　　AVERRHOACEAE
　　　　　　　　　　　　　　　　　　　　(formerly OXALIDACEAE)
CARAMBOLA—Native to Malaysia and common throughout tropical Asia, the
East Indies and the Pacific islands. Tree, to 30'; leaves evergreen, compound, with 3
to 7 pairs of oval, pointed, smooth leaflets, 1 ½ to 3" long; flowers rose-purple, 5-
petaled, in small clusters; fruit 3 to 5" long, oblong or oval with 3 to 5 prominent
angles, a cross-section of the 5-angled fruit suggesting a star; skin waxy, smooth, thin,
orange-yellow; flesh yellow, crisp, juicy, acid to subacid, muskily aromatic, edible
raw or cooked. Propagated by seed, grafts or air-layers; blooms and fruits off and on
throughout the year.

Avicennia germinans Stearn (*A. nitida* Jacq.)　　　　AVICENNIACEAE
　　　　　　　　　　　　　　　　　　　(formerly VERBENACEAE)
BLACK MANGROVE; HONEY MANGROVE; SALT-BUSH—Native to
southern U.S. and tropical America. Tree, to 70'; sprawling, irregular top; bark
maroon-brown, orange-red within; leaves narrow, up to 4" long, downy-white
beneath; flowers (spring) ½ " across, in clusters, white, fragrant, attract bees and
produce abundance of honey; fruit green, velvety, flat, 1" long. Grows on swampy
coasts and its foliage is sometimes coated with salt which can be collected for culinary
use. Sprouting seeds edible if cooked, but toxic raw. Hundreds of quill-like breathing
roots extend up through the ground around the tree.

Bamboo (*Arundinaria* spp., *Bambusa* spp.,　　　　　GRAMINEAE
　Dendrocalamus spp., *Phyllostachys* spp.)
Most ornamental bamboos are native to southern Asia. Several types may be seen in
South Florida, ranging from dwarf species with finger-thick canes to the giant bam-
boos, with canes 4" or more in diameter and to 70' or more in height. The HEDGE
BAMBOO, *Bambusa multiplex* Raeusch., which grows to 40', is one of the most at-
tractive and there are several cultivars, including the Stripestem Fernleaf, which can
be clipped to make a very dense, handsome growth. The YELLOW BAMBOO,
Phyllostachys aurea. Riv., up to 30' and one of the principal sources of fishing poles,
is not as well adapted to South Florida as it is further north, its range extending to

Maryland. The most frequently seen of the giant species is the COMMON, or FEATHERY, BAMBOO, *Bambusa vulgaris* Schrad., the shoots of which are eaten. This is grown to some extent in the Okeechobee area for furniture, lamp bases, etc., though it is considered inferior to the imported material. Its variety *vittata* A. & C. Riv. (var. *aureo-variegata* Hort.) is the GOLDEN BAMBOO with yellow stems striped with green.

Barleria cristata Linn.   ACANTHACEAE
BLUEBELL BARLERIA; PHILIPPINE VIOLET—Native to India and East Indies. Shrub, to 6', erect branches with yellowish down; leaves oval, up to 4" long, slightly rough-hairy; flowers (especially in fall) 5-lobed, flaring at end of 2 to 2½" tube, lavender-blue, pink, white, or white with lavender stripes, in showy spikes, each flower clasped by a pair of spiny bracts. Grows readily from seed or cuttings. The white-flowered type runs wild in South Florida.

Bauhinia blakeana Dunn.   LEGUMINOSAE
HONG KONG BAUHINIA—Native to southeast Asia. Tree, to 50 ft. tall, bushy and spreading; leaves nearly evergreen, 2-lobed, to 8" wide; flowers orchid-like, to 6" wide, rose-purple, in elongating sprays at the branch tips (from October to March); non-fruiting; must be propagated by cuttings, air-layering or grafting. Introduced from Hong Kong in 1953 by the University of Florida's Subtropical Experiment Station, Homestead, it has become a favorite ornamental. In 1958 it was adopted as the Official Flowering Tree of Boca Raton and in 1965 this species was made the national flower of Hong Kong.

Bauhinia galpinii N. E. Brown   LEGUMINOSAE
RED BAUHINIA—Native to tropical Africa. Shrub, climbing (by coiling branch tips) or sprawling and forming a mound, to 10 ft. high; leaves semi-deciduous, 2-lobed, to 3" wide; flowers nasturtium-like, 3" across, orange-red, in terminal clusters (from spring to late fall). Seed pods dark-brown, flat, to 5" long, with 4 to 7 seeds, regularly borne by some specimens, rarely by most. Propagation usually by air-layers. Thrives in full sun; is drought-resistant.

Bauhinia monandra Kurz   LEGUMINOSAE
BUTTERFLY-FLOWER; PINK BAUHINIA—Evidently native to Burma, though often reported to be a tropical American species; naturalized in West Indies. Shrub or tree, to 40'; leaves deciduous, divided into 2 lobes, pronouncedly butterfly-like, up to 8" long, underside light-green with a bloom; flowers up to 4" wide, light-pink with one yellowish or whitish petal streaked or flecked with purplish-red; fruit a pod up to 10" long containing glossy, black seeds. Fast-growing from seed. Blooms continuously except during winter when tree is bare.

Bauhinia purpurea Linn. (*B. triandra* Roxb.)   LEGUMINOSAE
ORCHID TREE; MOUNTAIN EBONY—Native to Asia. Tree, to 40' with slender trunk and drooping branches; leaves deciduous, 2-lobed, somewhat butterfly-shaped, up to 6" across; flowers to 5" in diameter, orchid-like, 3- or sometimes 4-stamened, red, pink, white or lavender, with non-overlapping petals to ¾" wide, appearing in

[30]

fall while tree is in foliage, followed by slender, foot-long pods which are edible when green, later turn brown and are unattractively conspicuous when tree sheds its leaves. Common in South Florida. Propagated by seed or grafting; grows rapidly. Pods split open with an audible snap.

Bauhinia tomentosa Linn.   LEGUMINOSAE
ST. THOMAS TREE; YELLOW BAUHINIA; BELL BAUHINIA—Native to India; naturalized in tropical Africa. Shrub, or small tree, to 25', with slender, drooping branches; leaves evergreen, 2-lobed, 1 to 3 ½" wide, downy beneath except in var. *glabra;* flowers (spring or fall or all summer) 2 to 4" long, never fully opening, yellow, usually with a maroon blotch at the base of one petal; pod 3 ½ to 5" long, to ¾" wide, flat, with 6 to 12 seeds. Flowers, pods, seeds, bark of tree and roots used in folk medicine. Bark yields fiber. Young leaves acid in flavor and eaten in the East Indies. In Hindu temples, on Dassera Day, the leaves are blessed and distributed as symbols of gold. Easily grown from seed. Introduced into Florida before 1900 but unfortunately not as well known as it should be.

Bauhinia variegata Linn.   LEGUMINOSAE
ORCHID TREE; BUDDHIST BAUHINIA—Native to India. Tree, to 40'; leaves deciduous, 2-lobed, butterfly-shaped, up to 5" wide; flowers orchid-like, up to 4" wide, 5-stamened, lavender or rose with red and yellow markings, with petals to 1 ¼" wide, overlapping, appearing in spring when tree is nearly devoid of leaves; pod up to 2' long, spirals as it springs open and can be used for decorative purposes. Variety *candida* Roxb. (syn. *B. alba* Buch-Ham.) resembles *B. variegata* but flowers are white with green veins. Both grow quickly from seeds; are common in South Florida.

Orchid Tree—*Bauhinia variegata*

Herald's Trumpet
Beaumontia grandiflora

Bishopwood—*Bischofia javanica*

Winter Begonia—*Begonia heracleifolia*

Lemon Bottlebrush—*Callistemon citrinus*

St. Thomas Tree
Bauhinia tomentosa

Ylang-Ylang
Cananga odorata

Dwarf Poinciana
Caesalpinia pulcherrima

Beaumontia grandiflora Wall. APOCYNACEAE

HERALD'S TRUMPET—Native to northeastern India. Vine, high-climbing, woody; leaves broad-oval, pointed, up to 9" long, dark-green and glossy on top, conspicuously veined; flowers (spring) white, to 6" long, trumpet-like, the flare 3" wide, 5-pointed and recurved, the tube clasped at the base by 5 sepals an inch or more in length; borne in large clusters, fragrant. Grown most easily from root cuttings in full sun or partial shade; fast-growing and heavy; requires strong support. Bark used for fiber.

Begonia heracleifolia Cham. & Schl. BEGONIACEAE
WINTER BEGONIA—Native to Mexico. Herbaceous plant with ridged, red-mottled, hairy leafstems up to 4' in height; leaves handsome, to 1' wide, hairy, deeply cut into as many as 9 points with purplish-red, irregular edges and conspicuous yellowish veins; flowers up to 1" across, pink or white, clustered at top of upright flower stalks. Blooms in winter. Fairly common in South Florida.

Beloperone guttata Brandg. (*Justicia brandegeana* Wass. & Sm.) ACANTHACEAE
SHRIMP PLANT—Native to Mexico. Herb or shrub, to 8', slender-stemmed, bushy; leaves somewhat tongue-shaped, pointed, to 5" long, downy; flowers in clustered spikes up to 4" long, white and protruding slightly from overlapping, yellowish to brownish-rose, heart-shaped bracts which are the showy feature of the plant. Effective and commonly used in mass plantings in South Florida. Cuttings root readily.

Billbergia pyramidalis Lindl. (*B. thyrsoidea* Mart.) BROMELIACEAE
Native to Brazil. Herb, to 2' with leaves arranged in tiered rosettes something like the crown of a pineapple; leaves light-green with spiny-toothed edges; flowers rose-colored, covered by 3"-long floral bracts in pyramidal central clusters; bracts red, sometimes tipped with purple. Self-multiplying; in sunny locations; may be grown on trees; throat must be kept filled with water.

Bischofia javanica Blume BISCHOFIACEAE
(formerly EUPHORBIACEAE)
BISHOPWOOD—Native to tropical Asia, Java and the Pacific islands. Tree, to 60 ft., with dense, rounded head; leaves alternate, compound, with 3 oval leaflets to 8" long, pointed and edged with fine teeth; flowers small, greenish-yellow, in sprays, male and female on separate trees; fruit round, 1/3" wide, brown, fleshy, containing 3 to 6 seeds. Thrives best and remains evergreen in moist soil; leaves turn red and fall during droughts. Fast-growing from seed or cuttings; a popular ornamental, now naturalized in South Florida and becoming a "weed" tree! Valued for timber in India.

Bixa orellana Linn. BIXACEAE
ANNATTO—Native to tropical America. Tree, to 30', bushy; leaves evergreen, up to 7" long, heart-shaped with pointed tip; flowers (in fall) like briar roses, 2" across,

pink, with deep-rose buds, in large clusters; fruit a hairy capsule up to 2½" long, brownish-green, maroon or bright-red in color, splits open and reveals small seeds coated with orange-red pulp. Pulp used to color rice, butter, cheese and to dye textiles. Grown from seed or cuttings in full sun; needs protection from wind.

Blighia sapida Kon. SAPINDACEAE
AKEE; VEGETABLE BRAIN—Native to West Africa. Tree, to 40', open-branched; leaves semi-deciduous, compound with bright-green leaflets to 6" long; flowers white, small, on hanging stems; fruit a bell-shaped, 3-lobed pod up to 4" long; yellowish, partly or largely blushed with bright-red; poisonous until fully ripe when pod naturally splits open in 3 parts each containing a firm ivory-colored meat (aril) attached to slender pink or red membrane and tipped with a large, black, shiny seed. When fresh and firm, the arils may be eaten raw but are commonly cooked and are exceedingly popular in Jamaica. Arils with undeveloped seeds should be avoided, also those which are overripe and soft. Fast-growing from seed but not recommended for dooryard planting. Seeds are always poisonous.

Bombax ceiba L. (*B. malabaricum* DC.) BOMBACACEAE
RED SILK-COTTON TREE; MALABAR SIMAL TREE—Native to East Indies, southern Asia and Australia. Tree, grows to large size, with numerous or few thick, pointed spines on trunk and branches; wide-spreading top; leaves deciduous, attractive, compound, divided into 5 to 7 pointed leaflets up to 6" long and spread like the fingers of a hand; during winter, the bare branches put forth a succession of beautiful, large, dark-red or orange-red, waxy flowers which are, incidentally, edible. Fruit is a 5-compartment pod up to 7" long, containing floss and small black seeds. May be grown from seed or air-layers, but large cuttings root easily and bloom early.

Bougainvillea glabra Choisy NYCTAGINACEAE
LESSER BOUGAINVILLEA—Native to Brazil. Shrub, to 10', or climbing; bushy, spiny; leaves tongue-shaped, up to 1½" long; flowers (all year) cream-colored and tiny but cupped in 3 petal-like bracts in tulip-like formation about 1" long and in showy clusters, magenta-red, purple, or white. Grown from cuttings in full sun. Purple type may be close-clipped to any form and still bloom plentifully. Variety *sanderiana* Hort. has crimson bracts, begins to bloom when very small; may grow to 20' or more.

Bougainvillea spectabilis Willd. NYCTAGINACEAE
Native to Brazil. Shrub, or climbing, to 25'; thorny, with strong stems and larger leaves than *B. glabra* and larger bracts, red, pink or coral, in large clusters. Variety *lateritia* Lem. has brick-red bracts.

Bougainvillea trolli Heimenl. NYCTAGINACEAE
TROLL'S BOUGAINVILLEA—Native to Bolivia. A variety, the Catalina Rose, is prized for its profuse clusters of light-pink to rose-colored bracts.

Boussingaultia leptostachya Moq. (*Anredera leptostachys* Steenis) BASELLACEAE
MADEIRA VINE—Native to Mexico, Central and South America, and West In-

dies. Vine, herbaceous with tuberous roots and fleshy, slender, twining, sometimes red, stems; leaves fleshy, bright-green, oval, pointed at both ends; flowers very small, in slender, pendant spikes, to 10″ long, white or greenish, fragrant. Common in uncultivated areas of Key West as an escape from gardens.

Breynia disticha Forst. (*B. nivosa* Small)　　　　　　　EUPHORBIACEAE
SNOWBUSH—Native to islands of South Pacific. Shrub, to 6′, with wiry, some-what zigzag, branches; leaves up to 2″ long, green with white patches, later all green; flowers greenish, small, in short clusters; fruit up to ½″ across, red. Often used for hedges. Variety *roseo-picta* Hort., the ROSE SNOWBUSH, also called JACOB'S COAT, has leaves variegated green, white, pink and red. Self-multiplying by suckers which transplant readily.

Brunfelsia americana Linn.　　　　　　　　　　　SOLANACEAE
FRANCISCAN RAIN TREE; LADY-OF-THE-NIGHT—Native to tropical America and West Indies. Shrub, to 8 or 10′ with slim branches; leaves evergreen up to 4″ long; flowers like tiny trumpets, the petals flaring at the end of a 4 or 5″ slender tube; white at first, turning yellow with age, very fragrant at night, borne singly or a few in a cluster; fruit round; ½″ across, orange or yellow. Common in Key West, occasional on the mainland of South Florida. Grown from seed or cuttings. Has been replaced in nursery trade by the similar *B. nitida* Benth. which has proved more satisfactory.

Bucida buceras Linn.　　　　　　　　　　　COMBRETACEAE
BLACK OLIVE; OXHORN BUCIDA—Native to the Bahamas, West Indies and Central America and possibly the Florida Keys. Tree, to 80′, with erect branches drooping at the ends and sometimes with short spines among the leaves; leaves, brief-ly deciduous, up to 3½″ long, clustered at the ends of the twigs; flowers small, greenish-white and without petals, in dangling spikes; fruit pointed-oval, curved, downy, about ⅜″ long, often transformed by mite damage into a peculiar gall. Propagated by seeds or air-layers; fairly slow-growing. Bark used in tanning; wood hard, valued for construction. Increasingly planted as a street tree in Greater Miami. Leaves turn orange or scarlet before shedding and are quickly replaced. (See photo at right.)

Bucida spinosa Jennings　　COMBRETACEAE
SPINY BLACK OLIVE—Native to the Bahamas and Cuba. Tree, to 25′, with slender, horizontal branches and zigzag twigs terminating in spines usually in 3's; leaves pear-shaped, clustered, to 1″ long; flowers minute, greenish, in small clusters; fruit black, oval, 1/6″ long. Grown from seeds and selected types grafted on *B. buceras*. Ideal for dwarfing in pots. This species and hy-brids with *B. buceras* popularized in South Florida in re-cent years, especially through the efforts of Herbert Early, nurseryman.

Butterfly Bush
Buddleja madagascariensis

Jelly Palm—*Butia capitata*

Gumbo Limbo—*Bursera simaruba*

Buddleja madagascariensis Lam.   BUDDLEJACEAE
MADAGASCAR BUTTERFLY-BUSH; SMOKE BUSH—Native to Madagascar.
Shrub, to 20', with spreading, arching branches, woolly twigs; leaves evergreen,
pointed, up to 6" long, dark-green and smooth on top, downy-white or -yellow
beneath; flowers orange, cupped, woolly on outside, in large, showy clusters. Leaves
used as a soap substitute; plant has medicinal uses in respiratory ailments. Grown
from cuttings.

Bursera simaruba Sarg. (*Elaphrium simaruba* Rose)   BURSERACEAE
GUMBO LIMBO; WEST INDIAN BIRCH; GUM-ELEMI—Native to South
Florida, the Bahamas, West Indies, Central America and Mexico. Tree, to 60' with
conspicuous light-bronze bark, flaking in thin scales; smooth, rounded trunk
sometimes forked into two upright branches which bring to thought a huge, up-ended
human torso. Leaves deciduous, clustered, pinnate with glossy green leaflets up to 3"
long; flowers greenish, small, in spikes; fruit angled, ½" long, dark-red, splits open.
Large cuttings often planted as living fenceposts in West Indies. Fast-growing. Tree
yields medicinal resin.

Butea frondosa Koen. ex Roxb. (*B. monosperma* Kuntze)   LEGUMINOSAE
FLAME-OF-THE-FOREST, or PARROT TREE—Native to India and Pakistan.
Tree, to 40' with rough, gray bark, crooked trunk and upright branches; leaves
deciduous, compound, with 3 broad-oval, leathery leaflets to 8" long, downy and
bronze when young, dull and ugly when old; flowers exceedingly beautiful, massed
along the ends of the branches from late January to March when the tree has few, if
any, leaves; calyx greenish-black, velvety, contrasting with the silky, orange, recurved petals, to 2" long. Seed pod flat, 4 to 8" long, containing 1 flat, brown seed
1½" long. Slow-growing from seed. Salt-tolerant; needs rich soil, neither too wet
nor too dry.

Butia capitata Becc.   PALMAE
BRAZILIAN BUTIA PALM; PINDO PALM; JELLY PALM—Native to Brazil.
Palm tree, to 20'; trunk up to 1½' thick; leaves grayish-green, feather-shaped,
arching; leafstems spiny; fruit in long clusters, nearly round, about 1" across, yellow,
fleshy but with large seed; used for jelly. Rare in South Florida but hardy as far north
as Virginia. Seeds germinate in 3½ months.

Caesalpinia crista Linn. (*Guilandina crista* Small)   LEGUMINOSAE
NICKER NUT; GRAY NICKERS; MOLUCCA BEAN; SEA BEAN; FEVER
NUT—Native to South Florida, Bermuda, the Bahamas, West Indies, tropical
America and Old World tropics. Vine, climbing or creeping, to 20' long; viciously
sharp, hooked spines on stems and underside of leaves; leaves twice-pinnate, to 1½'
long, with oblong or oval leaflets to 2½" long; flowers yellow, 5-petaled, in hanging
clusters up to 1' long; fruit a pod to 4" long and 2" wide covered with spines and with
spiny protruding tip, reddish-brown when ripe and dry, splits with open side upward,
holding the (usually 2) seeds like birds' eggs in a nest; seeds round-oval, hard, bluish-
gray, smooth, to ¾" long, do not fall until the pod disintegrates or they are shaken

out by high winds or other disturbance; since the vine grows near the seashore, they are carried by ocean currents. Kernels contain quinine-substitute used in folk medicine. Vine is aggressive, rapidly increasing and overwhelming other vegetation on Key Biscayne.

Caesalpinia gilliesii Wall. (*Poinciana gilliesii* Hook.)   LEGUMINOSAE
PARADISE POINCIANA; also called "BIRD OF PARADISE"—Native to South America. Shrub or small tree, spineless, with evergreen, feathery foliage; leaflets to ½" long with black dots near edge on underside; flowers (summer) in erect, elongated clusters, yellow with protruding red stamens to 5" long; fruit a pod 4 " long, ¾" wide. Grown from seed.

Caesalpinia pulcherrima Swartz (*Poinciana pulcherrima* Linn.)   LEGUMINOSAE
DWARF POINCIANA; BARBADOS PRIDE; FLOWERFENCE—Found in tropics of both hemispheres. Shrub, to 15', open-branched, slightly spiny; foliage evergreen in South Florida, deciduous further north; feathery, soft, bright-green; flowers (nearly all year) in erect, pyramidal clusters, brilliant scarlet-and-yellow, with long, red stamens; fruit a pod to 4" long and ¾" wide. Variety *flava* has yellow flowers. Immature seeds edible; leaves used as fish poison; flowers and roots yield dye; all parts used medicinally. Grown from seed (pre-soaked or scarified to hasten germination) in full sun; drought- and salt-tolerant.

Cajanus cajan Millsp. (*C. indicus* Spreng.)   LEGUMINOSAE
PIGEON PEA—Probably native to Africa. Shrub, to 10', may live 4 years; bushy, with very slender branches; leaves deciduous, compound, with leaflets to 4" long; flowers like those of the common pea, yellow, or variegated yellow and dark-red; fruit a green or green-and-red pod up to 3" long; contains plain or spotted peas commonly eaten in the tropics. These are the "peas" of the "no peas, no rice, no coconut oil" in the Bahamian song. (See photo on p. 40.)

Caladium biocolor Vent.   ARACEAE
FANCY-LEAVED CALADIUM—Native from the West Indies to Brazil. Herb, tuberous, perennial, dying back in winter; leaves, long-stalked, shaped like arrowhead, to 1' long, silky, thin, nearly translucent, handsomely dappled and streaked with red, pink, cream, yellow, light-or dark-green, silver or white. There are thousands of named cultivars. Dr. Henry Nehrling was a pioneer in hydridizing in Florida and Sebring is the national leader in commercial production of tubers. Propagated by division of tubers; planted in spring. Needs moisture and shade.

Calliandra haematocephala Hassk.   LEGUMINOSAE
RED POWDERPUFF—Native to tropical America but original site unknown. Shrub or tree to 25'; leaves evergreen, twice-compound, each main division having 5 to 10 pairs of oblong leaflets to 1 ½" long; flowers in dense heads 2 to 3" wide with showy, crimson stamens to 1 ¼" long; seed pod brown, flat, ridged. Fast-growing, winter-blooming; grown from seeds or cuttings.

Calliandra surinamensis Benth.   LEGUMINOSAE
PINK POWDERPUFF—Native to northeastern South America. Shrub or bushy
tree to 10' high; leaves evergreen, twice-compound, each main division having 8 to 12
pairs of leaflets ¾" long; flowers fragrant, in fluffy heads with pretty pink-and-white
stamens 2" long; blooming more or less continuously; seed pod dark-brown, flat with
thick edges, to 3" long. Grown from seed and suitable for dry locations.

Callicarpa americana Linn.   VERBENACEAE
AMERICAN BEAUTYBERRY—Native from South Florida to Virginia and Texas. Shrub, 3 to 7' high, with slender stems; leaves oval, pointed at both ends, to 6"
long, slightly rough, with fine-toothed edges; flowers mauve in clusters between the
leaves; fruits purple, round, ⅛" wide, massed along the stems, in autumn or nearly all
year with adequate moisture. One of the most conspicuous shrubs of pinelands and
hammocks. Fruiting stems useful for decoration. Grown from seed in partial shade.

Callistemon citrinus Skeels (syn. *C. lanceolatus* DC.)   MYRTACEAE
LEMON BOTTLEBRUSH—Native to Queensland. Shrub, usually to 15', up to 30'
in its native land; branches erect or arching; leaves evergreen, stiff, pointed, to 3"
long; flowers, with prominent bright-red stamens, borne in upright, 2- to 4-inch
spikes to 2½" wide. Variety *splendens*, introduced by Kew Gardens, is preferred to
the ordinary type. Less common in South Florida than further north in the State.
Often mistaken for *C. speciosus* BC., the SHOWY BOTTLEBRUSH, which is considered superior because of the longer flower spikes (to 6") and more conspicuous
gold tips on the stamens. Propagated by seed, cuttings and air-layers; fast-growing.

Callistemon viminalis Cheel.   MYRTACEAE
WEEPING BOTTLEBRUSH—Native to New South Wales. Tree, usually to 20' but
up to 60' in its native land; branches long and drooping; leaves evergreen, alternate,
slender, to 4" long, furry when young; flowers, with numerous deep-red stamens, in
long, terminal, cylindrical spikes 1½" wide; seed capsules contain tiny, papery seeds.
Fast-growing. Since distribution by U. S. Plant Introduction Station in 1956 has
become the most common bottlebrush in South Florida because of its great abundance of blooms, especially in spring.

Calonyction aculeatum House (*Ipomoea alba* Linn.) CONVOLVULACEAE
LARGE MOONFLOWER—Native to the tropics generally; found wild in South
Florida. Herb, with twining stems to 20', often prickly; leaves somewhat heart-
shaped, angled or 3-lobed, to 8" long; flowers (all year) flaring to a width of 6" at
end of slender 6" tube, white with greenish ridges radiating from center; blooms at
night, fragrant; fruit a seed capsule ¾" long. Milky sap of stems skin-irritating; used
to coagulate rubber; stems used to string tobacco leaves for drying; leaves edible,
cooked.

Calophyllum inophyllum Linn.   GUTTIFERAE
MAST-WOOD; KAMANI; incorrectly called ALEXANDRIAN LAUREL—Native
to tropical Asia. Tree, to 60'; leaves evergreen, dark, glossy, leathery, to 8" long and

Wild Cotton
Calotropis procera

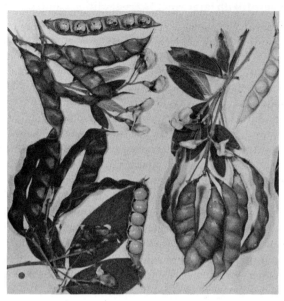

Pigeon Pea–*Cajanus cajan*

Bird Pepper–*Capsicum frutescens*

4" wide, with numerous, conspicuous parallel veins at right angles to midrib; blossoms white, about ¾" wide, in spikes, fragrant; fruit round, 1" across, yellow-skinned, with large round kernel within a thin, brittle shell. Nut-like kernel poisonous except for half-ripe endosperm which is eaten after pickling; mature kernel yields domba oil, an illuminant. Tree yields a gum, is valued for boat-building and cabinetwork; leaves, bark and roots have medicinal uses. Slow-growing from seed; ornamental and resistant to wind and salt; valued for coastal planting.

Calotropis procera R. Br.   ASCLEPIADACEAE
ST. THOMAS BUSH; WILD DOWN; WILD COTTON. (Also called GIANT MILKWEED but this latter name better applies to *C. gigantea* R. Br.)—Native to West Indies, tropical America and Old World tropics. Naturalized in the Bahamas. Shrub, to 15', with upturned branches bearing leaves in opposite pairs spaced along their entire length; leaves to 8" long, oval, with conspicuous white veins; young leaves velvety-white; flowers deep-lavender or white with reddish tint, 1" across, in clusters; fruit a pod up to 5" long containing numerous seeds, to each of which is attached a bit of silken floss used as a substitute for kapok. The plant produces a milky sap which is a skin-irritant to many people, contains calotropin, a heart-poison, but used medicinally and in cheese-making and in brewing beer. Leaves, roots and bark also medicinal.

Cananga odorata Hook.f. & Thoms.   ANNONACEAE
YLANG-YLANG—Native to Burma, Java and the Philippines. Tree, narrow, erect, to 40' or more, with downward-slanting branches; leaves evergreen, alternate, oblong-elliptic, to 10" long and 3" wide; flowers highly fragrant, especially at night, solitary or in a series of small clusters along the ends of the branches, greenish-yellow, changing to deep-yellow with age; petals, normally 6, ribbon-like, to 3" long; fruits oval, ½" to 1" long, black, fleshy, with 2 to 12 seeds. Oil distilled from the flowers abroad is prized in perfumes. Fast-growing from seed. Tree blooms when very young and more or less continuously all year.

Canna generalis Bailey   CANNACEAE
COMMON GARDEN CANNA—Cannas are mainly native to tropical America. Herb, with tuberous rootstock and erect stem to 5' or more, formed, like the banana stalk, of overlapping leaf-bases; leaves long-oval, pointed, up to 1½', variously hued, green to deep-purple; flowers clustered at top of central stalk, each bloom to 6" wide and richly hued—yellow, red, orange or variegated. There are several species of *Canna* in cultivation and hundreds of named cultivars, a number of which may be seen in South Florida gardens. Propagated by rootstocks.

Capsicum frutescens Linn.   SOLANACEAE
BIRD PEPPER; RED PEPPER; CAYENNE PEPPER—Native to tropical America. Shrub, to 8'; leaves up to 5" long; flowers ½" across, white or lavender; fruit round, conical, or elongated and pointed, ranges from ¼" to 1' long; color may change from white to yellow, then purple and finally red. Fruits usually "hot" to handle. Grown from seeds.

[41]

Papaya—*Carica papaya*

Carica papaya Linn.   CARICACEAE
PAPAYA; FRUTA BOMBA; LECHOSA—Native to tropical America. Herbaceous plant, to 20'; with long, nearly horizontal leafstalks each carrying at its tip an ornamental, deeply-lobed leaf up to 2' wide; flowers white, about 1" across; the male flowers in elongated, drooping clusters; fruit borne on trunk, varies from round to stout club-shaped, and from 6 to 15" in length; cooked as a vegetable when green; when ripe is yellow-skinned with yellow or orange-red flesh, musky, sweet, like melon; large central cavity lined with soft, black seeds, round, 3/16" across, sometimes eaten for their peppery flavor. Papain in milky latex of leaves and green fruit tenderizes meat and is used medicinally. Grown commercially in South Florida with increasing difficulty because of virus problems. Small-fruited plants may be seen growing wild as "escapes".

[42]

Karanda—*Carissa carandas*

Carissa carandas Linn.   APOCYNACEAE

KARANDA—Native to India. Shrub, sprawling, to 10' high, or climbing to treetops; and bearing sharp spines, in pairs; leaves evergreen, oval, leathery, glossy to 1 ½ " long; flowers fragrant, white or pinkish, star-like, 1" wide, in clusters of as many as a dozen; fruit oval or round, ⅝ to 1" long, in compact bunches, skin smooth, turns from wine-red to black and contains white, gummy latex; flesh dark-red, acid, juicy; seeds, usually 4, small, flat. Ripe fruit edible raw or cooked. Thrives in dry soil and full sun. Grown from seed or cuttings; superior types may be grafted on own rootstock.

Carissa macrocarpa A. DC. (*C. grandiflora* A. DC.)   APOCYNACEAE

CARISSA: NATAL PLUM—Native to South Africa. Shrub, to 10', bushy, with strong, sharp, forked thorns; leaves rounded-oval, to 3" long, stiff, dark-green; flowers 4- to 6-petaled, white, fragrant, up to 2½ " wide; fruit round or long-oval, up to 1¼ " wide and 2" long; skin bright-red, glossy; pulp rose-red with milky juice and with several flat, brown seeds in center. Fruit juicy, strawberry-flavored, eaten raw, cooked or preserved. The plant flowers and fruits almost continuously. Grown from cuttings; usually as a hedge in South Florida. Compact, dwarf types, especially "Boxwood Beauty" popularized in recent years for borders and ground covers constitute an important addition to our landscaping materials.

Tufted Fishtail Palm–*Caryota mitis*

Caryota mitis Lour.   PALMAE
TUFTED FISHTAIL PALM—Native to Burma and Malaya. Palm tree, to 25', with slender stem, growing in clumps; leaves to 9' long, light-green, thin, divided into many sections, each of which suggests a fish's tail; flower-cluster a foot or more long. Tree full-grown when it begins to bloom at top; flowering progresses down the trunk and tree dies at end of flowering period which may cover several years. Fruit ½" across, round, turns from red to nearly black, outer coat poisonous and its juice irritating to skin but seed kernel edible. Bud, or "cabbage" edible after cooking, though bitter. Grown from seeds (viable 4—6 weeks) or by transplanting suckers.

Caryota urens Linn.   PALMAE
TODDY FISHTAIL PALM; WINE PALM—Native to India and Malaya. Palm tree, to 80' or more; trunk up to 1 ½' in diameter; leaves may be 20' long and 12' wide, divided into "fishtail" sections, thicker and more strongly ribbed than those of *C. mitis*; inflorescence hanging, 4 to 12' in length; sap, induced to flow from inflorescence by cutting, is potable and is made into sugar and wine; fruit round, 1" across; red, with skin-irritating juice. Plant yields fiber and an edible starchy substance; young leaf-shoots are eaten. Grown from seeds which remain viable 4—6 weeks. This palm blooms progressively from the top to the base and then dies (in about 30 years).

[44]

Casasia clusiaefolia Urban   RUBIACEAE

SEVEN-YEAR APPLE—Native to seacoasts of South Florida, Bahamas and West Indies. Shrub, to 10', may be bushy or tree-like; leaves evergreen, oblong, to 6" long, leathery, glossy, in erect clusters at branch tips; flowers white, star-like, up to 1 ½" across, in upright clusters, very fragrant; fruit oval, about 3" long and 2" wide; green and hard when immature, it blackens and shrivels as it ripens, when the skin is tender and pulp is jelly-like, dark-brown, seedy and licorice-flavored; edible but not desirable; seems almost ever-present on tree. Favored by mocking-birds. Grown from seeds for coastal landscaping.

Casimiroa edulis Llave & Lex.   RUTACEAE

WHITE SAPOTE—Native to central Mexico. Tree, to 25' with a spread of 20'; leaves semi-evergreen, palmately compound, with 3 to 7 elliptic, pointed leaflets to 6" long; flowers inconspicuous, small, whitish, in small clusters; fruit (in spring) nearly round or oblate, to 3" wide, with smooth, thin skin, yellowish-green to golden-yellow, and soft, white to yellow, edible flesh, of sweet flavor accompanied by some bitterness. The large, white seeds are toxic. The WOOLLY-LEAVED WHITE SAPOTE (*C. tetrameria* Millsp.) has larger leaflets, white and velvety on the underside. Its fruits are usually larger (to 4" wide), irregular in form and often with gritty particles in the flesh. Fairly common in fruit tree collections in the past; seldom planted today. Propagated by seed and grafting; fast-growing.

Seven-year Apple—*Casasia clusiaefolia*

Candle Bush—*Cassia alata*

Cassia alata Linn.   LEGUMINOSAE
RINGWORM CASSIA; CURE-ALL; CANDLE
BUSH—Native to tropical America and West Indies.
Shrub, to 12', upright or sprawling; leaves pinnate with
oblong leaflets up to 7" long and 2" wide; flowers (late
fall and winter) 1" wide, rich-yellow, in erect, very compact, cylindrical clusters to 6" or more in length, much
like thick candles before the flowers open, when the buds
are shielded by yellow, overlapping, petal-like bracts;
fruit a thin, broadly-winged pod up to 6" long. Leaves,
seeds, bark and flowers used medicinally for skin dis-
eases; bark used in tanning. Grown from seeds or cuttings in full sun. Should be cut back severely after blooming.

Cassia bicapsularis Linn.   LEGUMINOSAE
Native to the Bahamas, West Indies, tropical America
and Mexico. Shrub, to 10' high, with flexible, sprawling
branches, sometimes climbing like a vine; leaves compound, with 2 to 5 pairs of oblong or oval leaflets to 1 ¼ "
long; flowers bright-yellow, ½" long, in clusters of 4 to
8; seed pod nearly cylindrical, ⅜ " thick, and to 6" long,
slightly indented between the many disk-like, shiny,
brown seeds imbedded in sweet pulp. Leaves are
purgative; nectar reportedly toxic to honeybees. Grows
readily from seed; blooms for several weeks in late fall
and early winter. Has escaped from cultivation and
become an attractive "weed."

Cassia biflora Linn.   LEGUMINOSAE
BUSHY SENNA; MOSQUITO BUSH; TWO-FLOWERED CASSIA—Native to
the Bahamas, West Indies and tropical America. Shrub or small tree, to 15'; leaves
pinnate, with leaflets up to 1 ¼ " long; flowers (fall to spring) golden-yellow, to 1"
across, in loose, flat clusters; fruit a flat, brown pod, up to 4" long and ½ " wide; seeds
oval, flat, glossy, brown, 3/16" long. Common along streets of Key West; very showy
in full bloom. Grown from seeds; loses some leaves in dry spells.

Cassia fistula Linn.   LEGUMINOSAE
GOLDEN SHOWER; PUDDING-PIPE TREE—Native to India. Tree, attains 30'
in height; leaves briefly deciduous, light-green, compound, with oval leaflets up to 8"
in length; flowers (June—August) 2" wide, bright-yellow in foot-long pendent
clusters, borne so profusely that the tree seems to radiate sunshine; fruit a dark-
brown, cylindrical pod about 1" thick and up to 2' in length, well-known in the drug
trade for the laxative property of the sweet pulp. Seeds used in novelties. Fast-
growing from seed. Fairly common ornamental in South Florida.

Cassia javanica Linn.   LEGUMINOSAE
PINK-AND-WHITE SHOWER, or APPLE-BLOSSOM SENNA—Native to Java
and Sumatra. Tree, to 60', with slender, drooping branches; young trunk and
branches spiny; leaves deciduous, compound, with 5 to 20 pairs of oval or oblong
leaflets, to 2 ¼ " long; flowers (in May-June or later) 2 ½ " wide, in large clusters; calyx
dark-red; petals red at first, fading to pale-pink; seed pod smooth, cylindrical, 8" to 2'
long, containing nearly round, reddish-brown seeds. Fast-growing from seed; needs
full sun. A hybrid between this species and *C. fistula* is called RAINBOW SHOWER
and has been adopted as the official tree of Honolulu. Its flowers, borne in great
profusion, are a blend of red, yellow and white.

Cassia siamea Lam.   LEGUMINOSAE
KASSOD TREE—Native to southern Asia and East Indies. Tree, to 40'; leaves
evergreen, pinnate with rather leathery leaflets up to 3" long; flowers brilliant-yellow,
½ " wide, in pyramidal clusters to 1 or 2' long; fruit a thin pod ½ " across and up to
10" long. Flowers used in curries; leaves and pods reported toxic to livestock. Fast-
growing from seed.

Cassia surattensis Burm. f.   LEGUMINOSAE
GLAUCOUS CASSIA—Native to southeast Asia, tropical Australia and the islands
of the South Pacific. Tree, to 25', with slender trunk and branches and bushy, rounded crown; leaves evergreen, compound, with 6 to 10 blunt-oval leaflets to 2" long;
flowers golden-yellow, to 2 ½ " wide, in short spikes; abundant in spring and again in
fall. Seed pods to 6" long and 2 ½ " wide, indented between seeds. May be propagated
by root shoots. A fine little tree, in recent years much planted in dooryards and parks.

Casuarina equisetifolia L. (*C. litorea* L.)   CASUARINACEAE
AUSTRALIAN PINE; BEEFWOOD—Native to Australia. Tree, somewhat pine-
like, of open growth, to 150'; technically leafless, its hair-like "needles" being fine,
jointed branches; small cones about ½ " broad and ¾ " long used in Christmas

Golden Shower—*Cassia fistula*

decorations and novelties. Wood red; bark used in medicine and in tanning. Fast-growing, salt-tolerant; seedlings often close-planted as a hedge or sea wall, trimmed to any desired height; individual specimens may be clipped round, square or more ornately. Commonly planted, also well-established as an escape from cultivation.

Casuarina glauca Sieber (formerly erroneously known as   CASUARINACEAE
     C. lepidophloia F.v.M.)
BRAZILIAN OAK; BLACK OAK; SCALYBARK BEEFWOOD—Native to Australia. Tree, to 70', dense, pine-like, not oak-like, with long, hanging "needles" giving it a soft, brushy appearance; suckers spring up around base; cone oblong or nearly round and up to 1" long, but unfamiliar in Florida. Propagated by suckers or grafted on *C. equisetifolia* seedlings. Commonly grown as a windbreak.

Catharanthus roseus Don (*Vinca rosea* Linn.;   APOCYNACEAE
     Lochnera rosea Reichb.)
MADAGASCAR PERIWINKLE; CAPE PERIWINKLE; OLD MAID—Probably tropical American; not native to Madagascar. Herb, upright, to 2' high with milky latex; leaves ovalish, to 3" long; flowers 5-lobed, to 1½" wide, at end of slim, 1" tube, lavender-pink, sometimes with a purplish-red center. Variety *alba* has white flowers; variety *oculata* is white with a red center. Fruit a V-form pair of slim, cylindrical seed pods, 1" long. Plant reported toxic to livestock. Has many uses in tropical folk-medicine; yields an alkaloid which has been employed since 1961 in treating leukemia. A favorite in planting bins; blooms constantly and spreads freely, forming attractive beds; is familiar as an escape in vacant lots.

Cecropia palmata Willd.   URTICACEAE
                                       (formerly MORACEAE)
SNAKEWOOD; TRUMPET TREE; SILVERLEAF PUMPWOOD—Native to lower West Indies. Tree, to 50'; slender trunk; umbrella top formed by open branches

Snakewood—*Cecropia peltata*

Glaucous Cassia—*Cassia surattensis*

Colville's Glory
Colvillea racemosa

Floss-Silk Tree
Chorisia speciosa

Java Glorybower
Clerodendrum speciosissimum

bearing at their ends large, decorative leaves very deeply divided into 7 to 11 lobes, conspicuously downy-white on underside; flowers in slender catkins; fruit compound, pencil-shaped, fleshy, about 6" long, light-green turning brownish when ripe, edible but insipid. Often planted as an ornamental in South Florida. Should not be confused with *C. peltata* Linn., native to the West Indies and northern South America, which attains 60' and has foot-wide leaves, equally handsome but divided only 1/3 of the distance to center into 7 to 11 lobes, with the leafstem attached near the center of the leaf; fruit about 3" long. The *Cecropia* trees are fast-growing and many harbor ants in their characteristically hollow branches and young trunks. The former are used for wind instruments and the latter for rafts and barrels; wood readily ignites from friction; bark used for medicine and cordage and rubber has been made from the scant latex. Fallen leaves, dry and curiously curled, are valued for arrangements. Propagated by seed or air-layers.

Ceiba pentandra Gaertn. (*Eriodendron anfractuosum* DC.)   BOMBACACEAE
SILK-COTTON TREE; KAPOK TREE—Native to tropics of Old and New World. Tree, grows to over 100' and attains great width with huge buttresses when old; trunk of young tree thorny; leaves compound, with whorl of 5 to 9 leaflets, each to 7" long, shed before blooming; flowers white or pink, 2 to 3" long in dense clusters; fruit a black pod, to 8" long, filled with brown seeds and cottony floss which blows about ground when pod opens. This floss is the kapok used in cushions and life-preservers. Fast-growing from seed.

Cephalanthus occidentalis Linn.   NAUCLEACEAE
(formerly RUBIACEAE)
BUTTONBUSH—Native to swamplands from south Florida to Mexico and eastern Canada; also Cuba. Shrub, or small tree, to 15' or even 30'; leaves deciduous, opposite or in whorls of 3, lance-shaped, to 7" long and 3½" wide; sometimes hairy on underside of veins; flowers (spring-fall) fragrant, white, tubular, with long stamens, in ball-like clusters to 1½" wide; fruit knobby, dark-brown, ¾" wide. Bark and leaves toxic to cattle and horses but young leaves eaten by deer.

Cereus peruvianus Haw.   CACTACEAE
HEDGE CACTUS; APPLE CACTUS—Native to South America. Cactus plant, to 50', columnar with upright branches, prominently ribbed or flanged, spiny, somewhat hairy; trunk and branches up to 8" across; flower night-blooming, white, to 6" across; fruit round-oval, to 3" or more long, red-skinned, with white pulp, juicy, faintly flavored, filled with tiny seeds. Grown from large cuttings.

Cestrum diurnum Linn.   SOLANACEAE
DAY CESTRUM; DAY JESSAMINE—Native to West Indies. Shrub, to 15'; leaves evergreen, glossy on top, up to 5" long and 1" wide; flowers tubular, ½" long, in hanging clusters, white, fragrant in daytime; fruit round or oval, violet at first, turning almost black, eaten by birds, toxic to humans in quantity.

Cestrum nocturnum Linn.   SOLANACEAE
NIGHT-BLOOMING JESSAMINE; POISON-BERRY—Native to West Indies. Shrub, to 10' or more, with slender, drooping branches; leaves pointed, 4 to 8" long and up to 1½" wide, glossy above and below; flowers tubular, ivory or yellowish, powerful fragrance at night often causes headache or other illness; fruit white, ½" across, poisonous, used in folk medicine.

Chamaedorea elegans Mart. (*Collinia elegans* Liebm.)   PALMAE
NEANTHE BELLA, or PARLOR PALM—Native to eastern and southern Mexico and Guatemala. Palm, slender, to 6'; leaves feather-like, with narrow leaflets to 8" long; may bloom when only 1' high; flowers tiny, yellowish, male and female in separate spikes; fruit round, black, ⅜" wide. A fast-growing, shade-loving palm, raised commercially in Florida for widespread distribution as a house plant. Seeds remain viable 4-6 weeks.

Chamadorea erumpens H.E. Moore   PALMAE
PACAYA—Native to Central America and Mexico. Palm, to 9' high, with several slender, bamboo-like stems in a cluster, bearing feathery leaves from the ground up; flowers fragrant, yellow, in branched cluster; fruit round, ¼" wide, turns from red to near-black. Seed germinates in about 7 months; palm is fast-growing in shady, moist situations. Popular in patios and in pots indoors. The GRASS-LEAVED PARLOR PALM, *C. seifrizii* Burret, from Mexico, usually less than 4' and either single-stemmed or clumped, is much grown, also.

Chiococca alba Hitch.   RUBIACEAE
SNOWBERRY; SNAKEROOT—Native to South Florida, the Bahamas, West Indies and tropical America. Shrub, upright, to 10', or climbing, sometimes to the top of tall trees; leaves up to 3" long, bright-green, smooth and glossy; flowers white, usually becoming yellow, bell-shaped, ⅜" long, in slender, pendent clusters, fragrant; fruit round, up to 5/16" across, white, in showy clusters. Conspicuous and beautiful on the edges of hammocks.

Snowberry—*Chiococca alba*

Floss Silk Tree—*Chorisia speciosa*

Chlorophytum capense Kuntze (*Anthericum elatum* Ait.)   LILIACEAE
SPIDER PLANT, or GRASS LILY—Native to South Africa. Herb, perennial, with horizontal rhizome and many fleshy roots; leaves, in rosette, grass-like, recurved, to 2' long and 1" wide, green or with white central stripe; flower stalk slender, 2 to 4' long, branched, drooping; flowers star-shaped, white, ½ to ⅝" wide, in loose cluster 6" to 1' long, followed by plantlets which may be left to dangle (from a pot or hanging basket) or transplanted. Propagated also by root division; grows rapidly in indirect light.

Chorisia speciosa St. Hil.   BOMBACACEAE
FLOSS-SILK TREE—Native to southern Brazil and Argentina. Tree, to 50', the thick trunk usually studded with short, stout spines till old, the branches long and spreading. Leaves, deciduous, palmately compound, with 6 to 7 slender, toothed leaflets; flowers variable in form and color, to 8" wide; 5-petaled, typically with prominent staminal tube; crimson, pink-and-white or pink with red streaks. Seed pod oblong, 3 to 6" wide, contains exceedingly soft and fine floss and numerous hard, dark-brown seeds like partly flattened peas. Fast-growing from seed. Seedlings may be grafted when very small. The tree is a glorious sight in full bloom in autumn. Floating floss from the exploded pods can be a nuisance if one has a heavily fruiting tree. Popularized since early 1950's when there were only a few old specimens in South Florida. The YELLOW-FLOWERED CHORISIA (*C. insignis* HBK) from Peru and Argentina, which may not be a distinct species, is less common.

Chrysalidocarpus lutescens H. Wendl. (*Areca lutescens* Bory)   PALMAE
CANE PALM; MADAGASCAR PALM; BAMBOO PALM (The old trade name, ARECA PALM, is obsolete).—Native to Madagascar. Palm tree, to 30', many-stemmed, forming bushy clumps; stems up to 6" thick, bamboo-like, ringed with golden-yellow; leaves feather-like, arched, with yellow stalk and midrib; leaflets 1"

[51]

wide, light-green; flowers in short clusters below the crownshaft and hidden by the leaves; fruit yellow when immature, dark-purple when ripe, oblong, ¾" long. Propagated by seed (viable 4-6 weeks) or suckers. One of the most popular palms of South Florida and often grown in pots for indoor use.

Chrysobalanus icaco Linn.

CHRYSOBALANACEAE
(formerly ROSACEAE)

COCOPLUM; ICACO—Native to South Florida, the Bahamas, West Indies and tropical America; possibly also tropical Africa. Shrub or tree, to 30', usually bushy; leaves up to 3½" long, rounded, leathery, glossy; new growth yellowish-green; flowers tiny, white, in spikes; fruit (late summer and fall) round or oval, up to 1¾" in diameter, yellowish-white, sometimes with pink cheek; skin soft; pulp white, cottony, sweet but mild; kernel of large seed edible and nutlike. Fruit eaten raw, canned in sirup or made into jelly. The inland cocoplum (*C. icaco* var. *pellocarpus* DC.), with smaller, dark-purple fruit, and reddish or yellowish new growth, has, in recent years, won recognition as an excellent ornamental and is available in nurseries. Slow-growing from seed; better propagated by air-layers or hardwood cuttings.

Chrysophyllum oliviforme Linn.

SAPOTACEAE

SATINLEAF; CAIMITILLO—Native to South Florida, the Bahamas, and West Indies. Tree, to 60', upright, with slender proportions; leaves up to 6" long, stiff, oval, pointed, glossy, dark-green above and glistening bronze-satin beneath; flowers white, tiny; fruit (spring) edible, oblong, up to 1¼" in length, dark-purple, with milky latex and sweet pulp. Grown from seed in sun or shade; salt-tolerant.

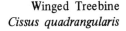

Lemon—*Citrus limon*

Winged Treebine
Cissus quadrangularis

Cissus quadrangularis Linn. (*Vitis quadrangularis* Wall.)   VITIDACEAE
WINGED TREEBINE—Native to tropical Asia and Africa. Climbing plant with fleshy, 4-sided stems, jointed at 2 to 4" intervals; leaves sparse, may be triangular or 3-lobed, up to 2" long with slightly toothed edges; flower buds rose-pink, flowers ¼" wide, light-pink inside, 4-petaled, in small clusters, 3 clusters on a short stem; fruit round, maroon, glossy, ⅝" wide, showy. Young shoots acid, used in curries; leaves and stems used as medicinal poultices; infusion has various medicinal uses for humans and livestock; fiber derived from stems and roots.

Citrus aurantifolia Swingle   RUTACEAE
LIME—Native to southern Asia. Tree, to 15', spiny; leaves evergreen, to 3" long, aromatic; fruit round or plump-oval, 1 to 3" long; skin green when unripe as commonly used, yellow when ripe and past commercial prime. Small, round variety is the Key, West Indian or Mexican lime, grown mainly on the Keys, where Key lime pie is a featured dessert. The large and more oval variety is the hardier Persian or Tahiti lime, grown commercially in South Florida.

Citrus aurantium Linn.   RUTACEAE
SOUR ORANGE; SEVILLE ORANGE—Native to southern Asia. Tree, to 3', spiny; leaves evergreen, up to 4" long, usually with winged petioles; flowers white, fragrant; fruit round, somewhat flattened; peel orange-yellow, rough; pulp orange-colored, acid. Excellent for marmalade, sometimes used for juice; introduced into Florida by early Spaniards; has been much used as stock on which to graft other *Citrus* trees.

Citrus grandis Osbeck (*C. maxima* Merr.)   RUTACEAE
PUMMELO; SHADDOCK—Native to southern Asia. Tree, to 30', sometimes spiny, with compact, rounded head; leaves evergreen, to 8" long; flowers white, clustered; fruit round, oval or pear-shaped, up to 9" across, with light-greenish-yellow, loose peel ½" or more thick; pulp composed of large juice-sacs, greenish-yellow or pink; may be more acid or sweeter than grapefruit. Peel may be candied. Rare in South Florida.

Citrus limon Burm. (*C. limonia* Osbeck)   RUTACEAE
LEMON—Native to southern Asia. Tree, to 20', of slender proportions, thorny; leaves evergreen, narrow, pointed, aromatic; flowers white, tinged with pink or dark-red on underside. The smooth, egg-sized "California lemon" is not often grown in South Florida but has recently become a commercial crop in the central part of the State. The larger ROUGH LEMON (*C. jambhiri* Lush.) is grown for home use and as a stock for other *Citrus*; the giant PONDEROSA LEMON, about 6" long and 4½" broad, with a knobby surface and pulp of indifferent flavor, is grown as a curiosity and the peel is used for candying. The dwarf MEYER LEMON (*C. meyeri* Tanaka), with smooth, rounded fruits, is nearly thornless, everbearing and useful in home gardens.

Citrus madurensis Lour. cv. Calamondin   RUTACEAE
  (formerly *C. mitis* Blanco)
CALAMONDIN ORANGE; PANAMA ORANGE—Originated in the Philippine
Islands. Tree, to 10' with very few thorns; leaves evergreen, up to 3" long, dark-green,
glossy; flowers white, small; fruit round, flattened, up to 1¾" wide; skin golden-
orange, loose, fairly thin; pulp highly acid but with fine lime-orange flavor, very juicy.
Fruit preserves well, makes good marmalade; juice excellent for ade, pie and ices.
Tree very attractive when fruiting and is a popular dooryard ornamental. Since 1960
has been grown from cuttings on a large scale for sale as a house plant under the trade
name of "Miniature Orange".

Citrus paradisi Macf.   RUTACEAE
GRAPEFRUIT—First known as a sport of the Pummelo in the West Indies. Tree,
to 50', upright, with smooth, brownish trunk; leaves evergreen about 6" long, dark-
green, leathery, shiny; flowers white, up to 1½" wide, clustered, fragrant; fruit round
or slightly pear-shaped, to 5" wide, few or many in a cluster; skin light-yellow; flesh
light-yellow, pink or deep-rose, acid to nearly sweet, juicy; may have 100 or more
seeds or be virtually seedless.

Citrus reticulata Blanco (*C. nobilis* var. *deliciosa* Sw.)   RUTACEAE
MANDARIN ORANGE; KID-GLOVE ORANGE—Native to southeastern Asia.
Tree, to 12 or 15' with compact head of small, spineless or slightly thorny branches;
leaves evergreen, narrow, up to 1½" long; flowers white, small; fruit rounded but
with flat ends, up to 3" across, rich-orange or reddish-yellow in color; skin loose, fairly thin; pulp subacid, of rich flavor. Fruit muskily aromatic. The Dancy Tangerine is
the leading variety. The Cleopatra Tangerine is considered one of the best rootstocks
for *Citrus* in South Florida. Tangelos (tangerine-grapefruit hybrids) are prized in
home gardens.

Citrus sinensis Osbeck   RUTACEAE
SWEET ORANGE—Native to China. Tree, to 40', with densely branched, conical
or rounded head, spineless or with few thorns; leaves evergreen, oval, to 4" long,
glossy; flowers white, more than 1" wide, very fragrant; fruit round, up to 3½"
across, yellow to golden-orange, borne singly or in clusters. Valencia, Jaffa, Pineapple, Parson Brown, and Temple, the latter a hybrid, are among the varieties commonly grown.

Clerodendrum indicum Kuntze (*C. siphonanthus* R. Br.)   VERBENACEAE
TUBE-FLOWER; TURK'S TURBAN—Native to eastern India. Shrub with un-
branched stems rising to 10'; leaves evergreen, narrow, lance-shaped, up to 8" long;
flowers white or cream-colored, slender, tubular, 4" long, flaring to 1½" wide, 5-
lobed, borne in large clusters at the ends of the stems; fruit round to ½" wide, green
at first, turning blue-purple, centered in conspicuous fleshy, red, star-shapd calyx.
Self-multiplying by suckers and runs wild. *C. minahassae* Teijsm. & Binn. is more
tree-like and its star-shaped calyx is twice as large and showy.

Pagoda Flower—*Clerodendrum paniculatum*

Clerodendrum paniculatum Linn.   VERBENACEAE
PAGODA FLOWER—Native to southeast Asia and Java. Shrub, to 6'; leaves evergreen, heart-shaped, 3- to 5-lobed, with indented veins, to 1' wide; flowers (June-October), star-like, yellow-and-red, ½" wide, borne in a striking, pyramidal inflorescence to 1 ½' high. First Florida specimens grown from cuttings obtained by the author from Charles Pennock's North South Nursery, Rio Piedras, Puerto Rico, in November, 1954. Edwin Menninger received seeds from Singapore in 1954 and Robert Rands introduced cuttings from Liberia in June, 1955. Plant thrives in full sun or partial shade; produces root shoots which may be transplanted.

Clerodendrum philippinum Schauer   VERBENACEAE
   (*Clerodendrum fragrans* var. *pleniflorum* Schauer)
FRAGRANT GLORYBOWER; CHRISTMAS ROSE; JAPANESE ROSE—Native to China and Japan. Somewhat woody plant, up to 8'; leaves evergreen, to 10" long, broad, pointed, tooth-edged, odorous; flowers white or pink, double, 1" wide, in flat clusters up to 4" across, very fragrant. Self-multiplying by root suckers; naturalized in South Florida, the West Indies and other warm areas.

Clerodendrum speciosissimum Van Geert (*C. fallax* Lindl.;   VERBENACEAE
   C. squamatum Vahl.)
JAVA GLORYBOWER (Incorrectly called Pagoda Flower, which is properly *C. paniculatum,* q.v.)—Native to Java. Shrub, up to 12', erect; leaves evergreen, heart-shaped, woolly, up to 1' long, on hairy stems; flowers bright orange-red, flaring to 2" wide at end of 1" tube, in upright clusters to 1½' long. Flourishes in shade; spreads like a weed from root suckers.

Clerodendrum speciosum D'Ombrain   VERBENACEAE
Shrub, semi-climbing; a hybrid between *C. thomsonae* and *C. splendens;* bears dark-green, glossy, evergreen leaves and large clusters of rose or blood-red flowers with cupped, star-shaped, purplish-red calyces which remain after the flowers are gone and are themselves highly decorative.

Clerodendrum thomsonae Balf. f.   VERBENACEAE
BLEEDING-HEART—Native to West Africa. Shrub, climbing and twining; leaves evergreen, broad-oval, pointed, up to 6" long; flowers, which protrude from loose clusters of white, inflated, 5-sided calyces about ¾" long, are blood-red, 5-petaled and ½" across with whitish stamens extending an inch beyond the petals. Summer-blooming; grown from cuttings; ideal for sunny locations where space is limited.

Clitoria ternatea Linn.   LEGUMINOSAE
BUTTERFLY PEA—Native to the Moluccas. Vine, herbaceous, slender, climbing to 20'; leaves alternate, compound, with 4 to 7 elliptic or oval leaflets to 1½" long;

Bleeding-Heart—*Clerodendrum thomsoniae*

flowers like those of the sweet pea, solitary, to 2" long, sometimes white or lavender, usually royal-blue with orange or white center; some plants have double flowers; seed pod flat, 3 to 5" long, splits down both sides, contains 5 to 12 brown, ½-inch seeds. Fast-growing from seed. Perennial and spontaneous in tropical climates; raised as a biennial in greenhouses or as an annual outdoors in the North.

Clusia rosea Jacq.   GUTTIFERAE
PITCH APPLE; MONKEY APPLE; COPEY; also called BALSAM APPLE, a name better limited to *Momordica balsamina* Linn., to which it is almost universally applied—Native to the Bahamas, West Indies and tropical America. (Has been reported in uncultivated areas on two Florida Keys but its nativity there is questionable.) Tree, to 50', with wide-spreading horizontal branches; leaves evergreen, broad-oval, to 8" long and 4½" wide, leathery, stiff; flowers attractive, pink or white and 3" across; fruit round, up to 3" wide, greenish-white with large pinkish calyx at stem end; suggests an unripe mangosteen but rind is divided into segments; turns black and woody and splits open to release seeds. Tree has far-reaching roots and, like the Strangler Fig, may grow on and encompass another tree. Leaves used by early Spaniards in lieu of writing-paper and also for playing-cards; leaves, bark and fruit have medicinal properties. Grown from seed or cuttings; often in planters where it soon exceeds the limited space.

Coccoloba diversifolia Jacq. (*C. floridana* Meissn.)   POLYGONACEAE
PIGEON PLUM; DOVE PLUM—Native to South Florida, the Bahamas, West Indies and northern South America. Tree, upright, to 70'; bark light-gray; leaves evergreen, oval, normally up to 4" long, leathery; flowers (spring) in slender spikes, whitish, abundant; fruit borne in slim, 3" long clusters, round, ¼ to ½" across, dark-purple, with one hard seed, edible. A splendid, compact tree, ideal for planting close to buildings and in parkways; should be more commonly utilized. Grown from seed. Leaves are remarkably variable, those on new shoots often twice the size of leaves on mature branches (see photo on page 58).

Coccoloba uvifera Jacq.   POLYGONACEAE
SEAGRAPE—Native to South Florida, the Bahamas, West Indies, coasts of Central America and northern South America. Tree, to 25' in height though often low-growing in clumps along beaches, branches spreading, beginning close to the ground and producing broad, rounded head if tree is not crowded; leaves circular, 4 to 8" across, red-veined, stiff; when old, turn red a few at a time; young leaves (in spring) a beautiful, silky bronze; flowers ivory, tiny, in spikes to 1' long; fruit in grape-like clusters, pear-shaped, up to ¾" long, slightly velvety, dark- or light-purple or almost white, with one pointed seed; excellent for jelly. Propagated by seed, cuttings or air-layers. Tree may be grown as a hedge and trimmed to shape. Leaves were used by early Spaniards as emergency "notepaper".

Coccothrinax argentata Bailey (*C. argentea* Sarg.)   PALMAE
FLORIDA SILVERPALM—Native to South Florida, the Bahamas, and West Indies. Palm tree, usually low-growing but may reach 40', with very slender trunk;

Pigeon Plum—*Coccoloba diversifolia*

leaves fan-like, almost circular, to 2' across, silvery-white on underside; flowers in clusters to 2' long; fruit round, up to ½" across, bright-purple at first, turning almost black, edible. Leaves woven into hats, baskets and matting. Seeds remain viable 2-3 months.

Cocculus laurifolius DC.   MENISPERMACEAE
LAUREL-LEAVED SNAILSEED—Native to southern Asia. Shrub, or small tree, to 15', with an equally broad head of drooping branches; leaves evergreen, lance-shaped, pointed, 3-nerved, leathery, shining, to 6" long; flowers and seeds unknown in Florida. Grown from cuttings as a specimen shrub or tall hedge, or for its handsome foliage which is long-lasting in wreaths. Leaves and bark contain the toxic alkaloid, *coclaurine.*

Cochlospermum vitifolium Willd. ex Spreng.   COCHLOSPERMACEAE
BUTTERCUP TREE, or WILD COTTON—Native to Mexico, Central America and northern South America. Tree, erect, to 75', soft-wooded, open-branched; leaves deciduous, alternate, 5- to 7-lobed, to 1' wide; flowers, in terminal clusters (January to May), bright-yellow, single with 5 petals or double, to 5" wide; seed pod velvety, oblate, to 3" wide, thin-shelled, containing "cotton" and kidney-shaped seeds 1/16" wide. Raised easily and rapidly from large cuttings. Drought-tolerant. Branches brittle and snap off in strong winds. In Central America, planted as a flowering fence, trimmed at the 6-ft. level.

Cocos nucifera Linn.   PALMAE
COCONUT—Nativity questionable, probably islands of Indian Ocean; formerly thought native to tropical America. Palm tree, to 100'; trunk slender, bulbous at base, often leaning or bowed, topped by rosette of feather-shaped leaves to 18' long and 6' wide, stiffly arched; flowers small, light-yellow, in cluster which emerges from canoe-shaped sheath which is often shellacked and used for flower arrangements and novelties; fruit to 15" long and 12" wide, composed of an oval, faintly 3-sided, thick, fibrous husk surrounding a nearly round nut with brittle, hairy shell containing ½" layer of white meat and refreshing coconut water. When immature, the meat is soft, jelly-like and eaten with a spoon or scooped out and made into an ice cream. It becomes firm before the husk has turned from green to brown, and only the expert can pick it at the jelly stage. Flowers and roots medicinal. The Golden and other Malayan varieties are disease-resistant and are being planted in South Florida and the Keys where lethal yellowing has killed so many ordinary coconuts. Ripe nuts will sprout in 3-6 months if heaped in the shade, hung up in trees and sprayed with water occasionally, or planted with the broad side up and showing just above the soil or mulch ánd watered every few days.

Codiaeum variegatum var. *pictum* Muell. Arg.   EUPHORBIACEAE
"CROTON"; VARIEGATED LEAFCROTON—Native to Malaya. Shrub, to 12', with short branches tipped with dense clusters of ornamental, evergreen, leathery leaves to 1' in length; broadly lance-shaped or deeply lobed, or narrow and twisted

spirally like a corkscrew, or extremely slim and straight. Colors vary from green with yellow spots or patches, to variegated green, red and yellow, or very deep purplish-red with pink markings. Flowers white, inconspicuous, small, in spikes. Stems contain somewhat milky, sticky juice which stains clothing. Roots and leaves medicinal; young leaves of some varieties edible when cooked. One of the commonest ornamentals in South Florida; easily propagated by cuttings and air-layers; thrives in sun or shade.

Coffea arabica Linn. RUBIACEAE
ARABIAN COFFEE; COMMON COFFEE—Native to southern and western Africa; introduced into Arabia. Shrub, to 15'; leaves to 6" long and 2" across, pointed, somewhat wavy, dark-green, shiny; flowers white, 4- or 5-petaled, in small clusters close to the branches, fragrant; fruit oval, cranberry-like, ½" long, dark-red, thin-fleshed, 2-seeded. The roasted seeds of this species are the principal coffee-beans of commerce. Rare but grown as a novelty in South Florida, with protection from severe cold.

Coleus blumei Benth. LABIATAE
COLEUS—Native to Java. Herb, to 3'; leaves soft, somewhat oval, pointed, tooth-edged, variegated with maroon, red, purple and yellow; flowers small, blue or white, in erect spikes. Variety *verschaffeltii* Lem. has rounded teeth and brighter-colored leaves, and is preferred as being more showy. Grown from cuttings; common garden and house plant.

Colocasia esculenta var. *fontanesii* A.F. Hill ARACEAE
PURPLE-STEMMED TARO—Native to Ceylon. Herb, with short, fleshy rhizome, from which rises a clump of elegant leaves; stems succulent, dark-reddish-purple, attached a few inches below the cleft; leaf-blade arrowhead-shaped, to 16" long, satiny, dark-green above, lighter below, violet-veined when young. An attractive plant for the edges of shallow ponds.

Colvillea racemosa Bojer LEGUMINOSAE
COLVILLE'S GLORY—Native to Madagascar. Tree, to 50' with ascending limbs and short side branches; leaves deciduous, twice-compound, finely divided, to 3' long; flowers (September-October) orange-scarlet, ⅝" to 2" long, 5-petaled with protruding stamens, in pendent, grape-like clusters 1½ to 2' long, massed at the tips of the branches. Leaves are shed after blooming. Seed pod nearly cylindrical, to 1' long. Grows slowly from seed; may bloom in 2 years.

Combretum grandiflorum G. Don COMBRETACEAE
SHOWY COMBRETUM—Native to Congo region of West Africa. Vine, woody, climbing or rambling; leaves slender-oval, pointed, up to 6" long; new growth bright-red; mature leaves often turn red at blooming time; flowers (winter) rich orange-red with yellow stamens, tubular, to 2" long, in numerous short-stemmed, one-sided clusters; fruits 5-winged, to 1½" long, in showy bunches turning from yellowish to red and then brown. Children pluck the individual flowers and suck the nectar from

the tube. Grown from seed or air-layers in full sun. Should be cut back severely after blooming.

Congea tomentosa Roxb.   SYMPHOREMATACEAE
                                                        (formerly VERBENACEAE)
WOOLLY CONGEA; SHOWER-OF-ORCHIDS—Native to Burma. Climbing shrub, woody, with numerous spreading branches; leaves deciduous, up to 7" long, oval with pointed tip, conspicuously veined, covered with minute hairs which make them rough to the touch; young shoots very hairy; flowers in long, loose sprays, white, tiny and nestled in the center of propeller-like clusters of 3 velvety, inch-long bracts, which are the showy features of the plant, the bracts being whitish at first (February), acquiring a pinkish-mauve or lavender hue over a period of many weeks. Grown from seed in full sun.

Conocarpus erectus Linn.   COMBRETACEAE
BUTTONWOOD; BUTTON MANGROVE—Native to South Florida, the Bahamas and West Indies. Shrub or tree, to 60', its trunk twisted or reclining on windy coasts; leaves evergreen, leathery, slender, pointed, up to 4" long, grayish-green; flowers tiny in rounded clusters grouped along an 8" spray; fruit like a small cone, ½" across, brown. Wood valued as timber and source of charcoal; bark medicinal and used in tanning. Variety *sericea,* the SILVER BUTTONWOOD, with attractive, light-gray foliage, has been adopted for general planting within the past 15 years. Grown from cuttings or air-layers.

Cordia sebestena Linn.   EHRETIACEAE
                                                        (formerly BORAGINACEAE)
GEIGER TREE—Native to Florida Keys, the Bahamas and West Indies. Tree, to 25', slender proportions; bark dark-brown, rough; leaves evergreen, broad-oval with pointed tip, rough, up to 8" long, in clusters on ends of branches; flowers up to 2" wide, brilliant orange-red in conspicuous flat clusters; "fruit" (a fleshy calyx), conical, to 1½" long, white, sweetish, edible, contains 1 or 2 brown seeds. Slow-growing from seed or air-layers. Named by Audubon after one John Geiger, a pilot and Key West "wrecker" of the 1830's.

Cordyline stricta Endl. (*Dracaena congesta* Hort.)   AGAVACEAE
Native to Australia. Herbaceous plant to 12'; resembles *C. terminalis* but leaves very slender, up to 2' long and range in color from green to rich-red and deep-purple; flowers lavender; fruit purple. Grown from seed or cuttings.

Cordyline terminalis Kunth. (*C. fruticosa* A. Chev.)   AGAVACEAE
TI—Native to southern Asia, East Indies and Polynesia. Woody plant, to 12' with few branches, upright, scarred, bare except at ends where leaves are snugly clustered like the "broom" of a feather-duster; leaves up to 2½' long and 5" wide, pointed; may be plain green, striped with white or red, or may be largely purplish, red or bronze; flowers small, in an erect spray up to 1' long, white, yellowish, lavender or rose; fruit red. Leaves used for thatching and garments and as food for livestock; root

Mexican Calabash–*Crescentia alata*

Ti–*Cordyline terminalis*

edible and used to make a potent liquor. Very common in Key West, and in early 1950's adopted into general use in South Florida landscaping. Cuttings and air-layers root quickly.

Cornutia grandiflora Schauer   VERBENACEAE
Native to Central America and southern Mexico. Shrub, or bushy tree, to 30', with quadrangular stems; leaves evergreen, strongly aromatic, broad-ovate, pointed, sometimes toothed, to 10" long, soft and downy; flowers fragrant, lavender, ¼" long, in erect, plump, pyramidal spikes to 15" long. Vigorous grower, requiring plenty of space.

Crescentia alata HBK. (*Parmentiera alata* Miers)   BIGNONIACEAE
MEXICAN CALABASH—Native to southern Mexico. Tree, to 45'; with erect, whiplike branches; leaves, clothing the branches, evergreen, compound, with usually 3, sometimes 5, oblong leaflets on a winged petiole giving the effect of a cross to 4" long; flowers bell-shaped, to 2½" long, yellow-green with purple stripes, emerging directly from the trunk and larger branches; fruit round, 2 to 5" wide, hard-shelled. The fruit shell is much used as a cup or flask for mescal. This tree has a special decorative flair suited to uncluttered modernistic settings. It is fast-growing from seed.

Crescentia cujete Linn.   BIGNONIACEAE
CALABASH TREE—Native to West Indies, Central and South America. Tree, to 40' with slender, open, spreading branches, trimmed along their entire length with closely-set, evergreen leaves up to 6" long; flowers greenish-yellow, purple-marked, to 3" long; fruit round or oval, green, hard-shelled, 4" to 1½' in length; pulp white, fibrous, edible but not desirable. Shell commonly used in West Indies as dipper or container. The tree is a noteworthy host of wild orchids and sections of calabash wood are often used by orchid fanciers as supports for their cultivated orchids. Fast-growing from seeds, but only occasionally seen. Succumbs to severe cold spells.

Crinum spp.   AMARYLLIDACEAE
CRINUM LILY—Herb, with onion-like bulb and upright, outward-curving, light-green, smooth leaves to 4' long and 3" wide; flower stalk upright, to 3' or more, with flowers clustered at top; flowers fragrant, up to 8" wide, funnel-shaped or divided into ribbon-like, curved segments. A number of species are commonly found in Florida gardens. The MILK-AND-WINE LILY, *C. sanderianum* Baker, native to Africa, has clusters of 3 to 6 flowers, dark-red outside, white inside, striped with red. The CEYLON CRINUM, *C. zeylanicum* Linn., native to tropical Africa and Asia, bears clusters of 10 to 20 blooms; buds dark-red, opened flowers white with red stripe; flower stalk purplish; many varieties. The CARIB CRINUM, *C. erubescens* Ait., native to tropical America, has clusters of 4 to 12 flowers, maroon outside and white with faint-pink stripe on the inside. The FLORIDA CRINUM, or FLORIDA SWAMP LILY, *C. americanum* Linn., a native of Florida and Georgia and west to Texas, bears clusters of 3 to 6 white flowers; often seen wild in the Everglades. *C. moorei* Hook. f. has handsome, bell-like, pink, rose or white blooms to 4" long. Propagated by division.

[63]

Crossandra infundibuliformis Nees.   ACANTHACEAE
CROSSANDRA—Native to India. Shrub, to 3'; leaves rich-green, glossy, to 5" long; flowers (March-May) salmon-colored, tubular with lobed lip 2" wide, borne in erect spikes above closely set bracts. Grown from seeds (4 in a capsule) or cuttings, in dense shade and moist acid soil.

Cryptostegia madagascariensis Bojer   PERIPLOCACEAE
(formerly ASCLEPIADACEAE)
MADAGASCAR RUBBER-VINE, erroneously called PURPLE ALLAMANDA— Native to Africa. Climbing shrub with strong, woody stems; leaves evergreen, oblong, up to 5" in length, leathery, dark-green, glossy; flowers deep-lavender, bell-shaped, to 3" wide; fruit a pointed, angled pod borne in pairs, end-to-end. *C. grandiflora* R. Br., the PALAY RUBBER-VINE, from India, has larger, pinkish flowers which fade to near-white; is rare in Florida but has run wild in the West Indies. Plants contain white latex used as a substitute for rubber; stems yield fiber; plants poisonous to stock and humans. Grown from seed, cuttings or air-layers.

Italian Cypress
Cupressus sempervirens

Cupressus sempervirens var.   CUPRESSACEAE
  stricta Ait.
ITALIAN CYPRESS—Native to eastern Mediterranean region. Tree, slender, columnar, to 100' or more, with short, compact branches from the ground up; leaves evergreen, scale-like, densely and closely set on shoots; cones nearly round, to 1 ¼ " long. Grown from cuttings or air-layers. An elegant conifer, prized for foundation planting, especially in restricted space but often without suitably formal surroundings.

Cycas circinalis Linn.   CYCADACEAE
CROZIER CYCAS; FERN PALM—Native to East Indies, East Africa and Guam. A cycad; a palm-like tree with stocky, rough-surfaced trunk up to 15', topped with a rosette of stiff, glossy, dark-green, feather-shaped leaves to 8' long and 2' wide. Some specimens have forked trunks with two or more heads. Male plant produces vertical, oblong, orange-colored odoriferous cone up to 2' high at the top of the trunk. Female cone is up to 1' high, broad and round-topped, its velvety, buff-colored sections opening as seeds form and standing erect, forming a crown, later bending outward and hanging down around trunk as the seeds mature. Seeds (rare without hand-pollination), orange-colored, oval, about 1 ½ " long, with a kernel which, after several soakings in water to remove poisonous principle, is made into a meal. Grown from seeds or offshoots. Fast-growing and popular in South Florida.

Geiger Tree
Cordia sebestena

Rubber Vine
Cryptostegia madagascariensis

Buttercup Tree
Cochlospermum vitifolium

Cat's Claw—*Doxantha unguis-cati*

"Croton"—*Codiaeum variegatum*

Umbrella Plant—*Cyperus alternifolius*

Cycas revoluta Thunb.   CYCADACEAE
SAGO CYCAS; FALSE SAGO PALM—Native to Java. A cycad; a palm-like tree
with thick, rough trunk to 10' high, with head of stiff, feather-shaped leaves to 7'
long, dark-green, glossy, edges of leaflets rolled under; leaflets narrower and shorter
than those of *C. circinalis* and more pointed. Male cone erect, cylindrical, up to 1 ½';
female cone rounded. Seeds to 1 ½" long, red, as many as 200 in bunch, with toxic
kernel. Edible farinaceous substance obtained from center of trunk. Grows from
seeds or offshoots. Slow-growing. Most common around older homesteads, especially in Redlands area of Dade County.

Cyperus alternifolius Linn.   CYPERACEAE
UMBRELLA PLANT—Native to Madagascar. Herb, aquatic, perennial, with thick
rhizome from which rises stiff, triangular stems to 6' high, topped by parasol-like
flowering heads composed of slender leaf-like bracts to 8" or 1' long surrounding a
central cluster of minute flowers in spikelets. Grown from seed, divisions or plantlets
developing in the inflorescence, and planted at pool edges or grown in pots.

Cyperus papyrus Linn.   CYPERACEAE
PAPYRUS—Herb, aquatic, perennial, with thick rhizome and erect, dark-green, 3-
angled stems to 10' or more, each topped by a large, circular brush of close-set,
thread-like, drooping bracts, with spikelets in the center. Thrives in shallow water or
wet land. Stem pith pressed into paper in ancient Egypt. Rhizome was roasted and
eaten.

Cyrtopodium punctatum Lindl.   ORCHIDACEAE
COWHORN ORCHID—Native to southern Florida, West Indies, Central America,
northern South America and southern Mexico. Orchid, epiphytic or terrestrial, with
thick, hornlike pseudobulbs 8" to 2' long; leaves, ribbon-like, to 30" long and 2"
wide, drooping at the tips; flowers frilled, 2" wide, yellow, purple and brown, in a
large branched panicle. Forms clumps to 3' broad. Wild plants formerly exploited,
now protected by State law.

[65]

Dalbergia sissoo Roxb.　　　　　　　　　　　　　　　　　　LEGUMINOSAE
SISSOO—Native to India. Tree, to 80', with open, somewhat spreading branches; leaves semi-evergreen, pinnate, leaflets 3 to 5, oval, pointed, to 3" long; flowers white or yellowish, small, in spikes, fragrant; fruit a slender, brownish pod up to 4" long. Wood valued as second only to teak in India; much used for furniture and carved objects; also yields a medicinal oil. Leaves used as fodder. Fast-growing from seeds.

Datura candida Pasq. (erroneously *D. arborea*).　　　　　　SOLANACEAE
ANGEL'S TRUMPET—Native to southern Mexico and Central America. Tree, to 15', with slender trunk; leaves oblong-oval, tapering to a pointed tip, soft, downy, to 16" long; flowers heavily fragrant, pendent, white, trumpet-like, to 12" long with spathe-like, single-pointed calyx; some forms have double blooms. Does not produce its smooth, cylindrical seed pods in Florida. Grows quickly from cuttings, also sends up shoots from the roots. Flowers abundant and beautiful, 2 or 3 times a year, but highly toxic, as is the fragrance to some people. This species and other woody relatives (including *D. suaveolens,* below) are transferred by some botanists to the genus *Brugmansia*.

Datura metel Linn. (*D. fastuosa* Linn.)　　　　　　　　　　SOLANACEAE
DEVIL'S TRUMPET; BLACK DATURA—Native to India. Herb, annual, to 5', downy; leaves to 8" long, smooth-edged or toothed; flowers trumpet-shaped, upright, white, yellow or purple, up to 7" long, often double; fruit a rounded, spiny pod, 1 ¼ " across with persistent, recurved calyx; fruit faces downward when mature. The so-called "*D. chlorantha*" is a fancy yellow form of this species. Grown from seed. All *Datura* species contain varying amounts of atropine, scopolamine and more or less hyoscyamine; have narcotic effects on humans and produce hallucinations or delirium.

Datura stramonium Linn.　　　　　　　　　　　　　　　　SOLANACEAE
JIMSON WEED; THORN APPLE—Native to tropics of Old and New World and found wild throughout much of eastern and southern U.S. Herb, annual, to 5'; leaves to 8" long, tooth-edged, odorous; flowers white or lavender, trumpet-shaped, up to 4" long and 2" wide, pointing upward; fruit, an erect, spiny, oval pod to 2" long, splits into 4 parts. Foliage and flowers used medicinally.

Devil's Trumpet—*Datura metel*　　　　Angel's Trumpet—*Datura suaveolens*

Princess Palm—*Dictyosperma album*

Datura suaveolens Humb. & Bonpl.   SOLANACEAE
    (erroneously *D. arborea* Linn.)
ANGEL'S TRUMPET—Native to Mexico. Shrub, to 15' with tree-like habit; leaves
evergreen, up to 1' long and 4" wide; flowers white, trumpet-like, up to 1' in length,
with tubular, 5-toothed calyx, nodding, may be single or double, often strikingly
abundant, fragrant, opening in the evening; fruit spindle-shaped, spineless, up to 5"
long, but absent in Florida. Easily grown from cuttings. This species less toxic than *D.
metel*.

Delonix regia Raf. (*Poinciana regia* Boj.)   LEGUMINOSAE
ROYAL POINCIANA; FLAMBOYANT TREE; FLAME TREE—Native to
Madagascar. Tree, to 40', with wide-spreading branches, gracefully arched top and
aggressive surface roots; leaves feather-like, to 2' long, bright-green, soft, with tiny
leaflets; shed during winter, leaving tree entirely bare; flowers (May-June) up to 4"
across, bright orange-red to deep-red with one white- or yellow-striped petal, are
borne in profuse clusters covering the tree with a riot of color; fruit a dark-brown,
strap-like pod, 2" broad and up to 2' long. Very fast-growing from seeds. Common in
South Florida as a dooryard and street tree. A yellow-flowered form, popular in
Kingston, Jamaica, has been successfully introduced.

Dictyosperma album H. Wendl. (*D. rubrum* Hort.)   PALMAE
PRINCESS PALM, or HURRICANE PALM—Native to the Mascarene Islands.
Palm, to 30' or even 50', with smooth, ringed trunk to 9" thick; leaves feather-like, to
12' long; often red-veined and with red-edged petiole when young; flowers (male and
female) pale-yellow in drooping cluster to 3' long; fruit purple, nearly round, ½"
long, one-seeded. Fast-growing from seed in shade or sun; highly wind-resistant.

Hondapara—*Dillenia indica*   Bitter Yam—*Dioscorea bulbifera*

Chamal—*Dioon spinulosum*

Dieffenbachia maculata Bunting (*D. picta* Lodd.)　　　　ARACEAE
DUMB CANE—Native to Brazil. Herb, with firm, fleshy stem to 8' tall and 1½"
thick; leaves oblong, tapering to a point at the tip, smooth, dark-green blotched with
white and pale-green or yellow. Common pot plant, easily raised from stem cuttings
which root readily in water. Sap watery, acrid, severely irritant externally and internally. There are numerous cultivars of this and related species.

Dillenia indica Linn.　　　　DILLENIACEAE
HONDAPARA—Native to India, Thailand and Malaya. Tree, to 60', with rounded
crown; leaves semi-deciduous, borne mostly at ends of branches, leathery, conspicuously veined, elliptic, toothed, to 14" long and 6" wide; flowers fragrant,
solitary, lovely, to 8" wide, with 5 white petals and a central mass of golden stamens
and styles; fruit glutinous and seedy, enclosed in thick, tough, fleshy, overlapping
sepals, yellowish and muskily odorous when ripe. The sepals, acid as rhubarb, are
eaten raw or cooked. Grows slowly from seed.

Dioon edule Lindl.　　　　CYCADACEAE
CHAMAL—Native to southern Mexico. Cycad, palm-like, with thick trunk to 7'
high, topped by a dense rosette of stiff, feather-shaped leaves to 5' long, with narrow,
closely set, sharp-pointed leaflets to 6" long. Male cone 1' or more long, whitish at
first, then dark-green. Female cone, on separate plant, low, rounded, covered with
brown fuzz; contains numerous chestnut-like, starchy seeds. *D. spinulosum* Dyer is
similar but bears sharp spines on the edges of the leaflets.

Dioscorea bulbifera Linn.　　　　DIOSCOREACEAE
BITTER YAM or AIR POTATO—Native to Old World Tropics. Vine, herbaceous,
perennial, twining, aggressively climbing and spreading; leaves alternate, heart-
shaped, glossy, attractively veined; flowers tiny in dangling clusters, male to 4½"
long, female to 9". Large subterranean tubers and aerial "bulblets" apt to be toxic;
sometimes edible after elaborate preparation and cooking. This vine is fast-growing
and ornamental but has escaped from cultivation and may shroud all the trees and
bushes in a vacant lot.

Diospyros digyna Jacq. (formerly *D. ebenaster* Retz.)　　　　EBENACEAE
BLACK SAPOTE—Native to Mexico and Central America. Tree, to 45', with short
trunk and dense, handsome crown; leaves evergreen, alternate, oblong, to 6" long
and 2" wide, smooth, leathery, glossy, dark-green; flowers inconspicuous, yellow-
green, male and female on same tree; fruit tomato-shaped, with leathery calyx at base
of stem, smooth, bright-green changing to olive-green when soft and ripe; flesh
brown, edible, sweet, mild in flavor; seeds brown, flattened, to ¾" long. Some fruits
seedless. A fine ornamental tree except for the nuisance of the fallen fruits, messy un-
derfoot. Fast-growing from seed.

Dipladenia splendens Hook. f. (*Mandevilla splendens* Woodson) APOCYNACEAE
PINK ALLAMANDA—Native to Brazil. Vine, herbaceous, twining, tuberous-
rooted; leaves semi-evergreen opposite, elliptic with long, tapering tip, heart-shaped
at base, to 8" long; flowers strikingly beautiful, tubular with 5 spreading lobes, to 5"

wide, rich pink with darker throat. Grown from cuttings. Of recent promotion by South Florida nurseries.

Dizygotheca elegantissima Viz. & Guill. ARALIACEAE
FALSE ARALIA—Native to the South Pacific islands. Shrub or small tree, with slender erect stem, to 10' or more; leaves evergreen, palmately compound, with 7 to 11 narrow, coarsely toothed, green or bronzed, drooping leaflets, 6 to 10" long. Grown from cuttings.

Dombeya sp. STERCULIACEAE
ROSEMOUND—Developed by Paul Soderholm, Research Horticulturist, U.S. Plant Introduction Station, Miami, from seed believed to be from Reunion. Shrub, to 6' with equal width, having 3-lobed leaves and large, showy clusters of pink flowers above the foliage. During November and December, the compact, rounded shrubs are covered with blooms busily visited by honeybees. Propagated by cuttings. Released to the nursery trade in June, 1967. Cuttings of two other admirable cultivars, "Perrine" and "Pink Clouds", distributed to nurseries in March, 1970.

Dombeya x *cayeuxii* Andre (formerly known as STERCULIACEAE
  D. wallichii Benth. & Hook.
PINK BALL—Native to Madagascar. Tree, to 30', often densely bushy; leaves evergreen, to 14" long and wide, heart-shaped with a point on either side, tooth-edged, soft-hairy, long-stalked; flowers (winter) pink, 1" across, in large, compact, rounded clusters suspended on hairy stems and partly hidden by the foliage; fruit a hairy 5-compartment seed capsule. Easily grown from seed or cuttings.

Doryalis hybrid. FLACOURTIACEAE
Accidental cross of *D. hebecarpa* (male) and *D. abyssinica* Warb. (female) at the U. S. Plant Introduction Station, Miami, yielded seed planted by the late Dr. R. Bruce Ledin at the University of Florida's Subtropical Experiment Station. The resultant shrub, having downy, deciduous leaves, long, sharp spines and reddish-orange, apricot-like, slightly velvety fruits, 1" or more across, attracted much attention. Propagated by cuttings and air-layers, this hybrid was very soon adopted by homeowners because of its tremendous crops of fruits at least twice a year. No vernacular name has been chosen for this plant.

Doryalis hebecarpa E. Mey. (*Dovyalis hebecarpa* Warb.) FLACOURTIACEAE
KETEMBILLA, or CEYLON GOOSEBERRY—Native to Ceylon. Shrub, to 15' high and twice as broad, more or less spiny. Leaves deciduous, alternate, oval, pointed, velvety, wavy, to 4" long; flowers small, greenish, male and female usually on separate plants but occasional specimens have perfect flowers; fruit (in winter) round, ⅝ to 1" wide, purplish-black, with light dots, velvety; pulp red-purple, juicy, sour, edible raw or preserved. Grown from seeds or cuttings.

[70]

Doxantha unguis-cati Rehd. (*Macfadyena unguis-cati* Gentry)   BIGNONIACEAE
CAT'S-CLAW—Native from the West Indies to Argentina. Vine, slender, climbing
to tree-tops; leaves evergreen, opposite, consisting of 2 wavy, elliptic, pointed,
3″-long leaflets and a 3-clawed tendril between them; flowers (in spring) rich-yellow,
funnelform, with 5 spreading lobes 3 to 4″ wide, borne profusely in clusters; seed
pod slender, flattened, to 1′ long. Grown from seeds, cuttings or layers.

Duranta repens Linn. (*D. plumieri* Jacq.)   VERBENACEAE
GOLDEN DEWDROP; SKY-FLOWER—Native to the Bahamas, West Indies,
Central America and northern South America, and also said to be native to South
Florida but this is doubtful. Shrub or small tree, to 18′ with slim, long-arching
branches; may be thorny; leaves evergreen, vary in form, up to 2″ long; flowers
lavender or white, in pendent clusters to 6″ long; fruit bright orange-yellow, round,
up to ½″ across, very showy. Fruits somewhat poisonous to humans and animals;
plant poisonous to livestock. Grown from seed or cuttings in full sun. Common ornamental in South Florida.

Ehretia microphylla Lam.   EHRETIACEAE (formerly BORAGINACEAE)
PHILIPPINE TEA; FALSE TEA—Native to China and East Indies. Tree, to 30′
with smooth, gray bark and slender branches, sometimes arching; leaves evergreen,
paddle-shaped, up to 2½″ long, clustered at ends of very short twigs, dark-green but
covered with dense, short, white hairs on top, light-green and less hairy on underside; flowers (all year) small, white or pinkish, in clusters; fruit round, ¼″ wide,
yellow or red turning to dark-blue on maturity, 4-seeded. Propagated by seed and
needs pruning. Leaves used as substitute for tea.

Elaeagnus philippensis Perr.   ELAEAGNACEAE
LINGARO—Native to the Philippine Islands. Shrub, climbing, or sprawling and
forming a mound to 10′ high and twice as broad; leaves evergreen, alternate, elliptic,
pointed, to 3″ long, glossy green above and silvery beneath; flowers (January-March)
fragrant, small, silvery outside, yellow inside, in small clusters among the leaves along
the ends of the stems; fruit oblong, to 1″ long, pink with minute, silvery scales, acid-
sweet, edible. Grown from seeds, cuttings or air-layers in full sun.

Enallagma cucurbitina Baill. (*E. latifolia* Small)   BIGNONIACEAE
BLACK CALABASH—Native to South Florida and West Indies. Tree, to 25′, upright with numerous erect, slender branches; leaves evergreen, oval, up to 6″ long,
tough, dark-green, glossy; flowers bell-shaped, to 2½″ long, lavender or yellowish;
fruit oval, up to 4½″ long, with thin, hard shell, inedible. Grown from seed. A fine,
dense, salt-tolerant ornamental but top-heavy and will blow over in windstorms.

Enterolobium cyclocarpum Griseb.   LEGUMINOSAE
EAR TREE—Native to the West Indies and tropical America. Tree, to 100′ or more,
with broad top and huge trunk; leaves deciduous, twice-compound, finely divided;
flowers tiny, white, in small, globose clusters; seed pod ornamental, dark-brown,
glossy, flat, nearly circular, fluted, to 4″ wide, containing many small, brown seeds.

[71]

Blue Sage—*Eranthemum pulchellum*

Fast-growing from seed; wind-resistant, with aggressive, far-reaching surface roots. Suitable only for large parks. Yields commercially valuable timber.

Epidendrum tampense Lindl. ORCHIDACEAE
BUTTERFLY ORCHID—Native to South Florida, Bahamas and Cuba. Orchid, grows on trees; leaves narrow, to 12" long, protruding from onion-like pseudobulbs; flower stalk to 3' long, branched; flowers up to 1½" across, greenish-yellow tinged with brown and with a white or yellow lip spotted with reddish-purple; fragrant. Propagated by division. **Needs** filtered sunlight or morning sun only.

Eranthemum pulchellum Andr. (*E. nervosum* R. Br.) ACANTHACEAE
BLUE SAGE—Native to India. Shrub, to 6'; leaves opposite, oval, pointed, often blunt-toothed, to 8" long; flowers in upright spikes to 3" long, medium- to dark-blue, tubular with 5 overlapping lobes spreading to ¾" wide; seed capsule oblong with 4 seeds. Grown from cuttings; needs shade and moisture.

Eriobotrya japonica Lindl. ROSACEAE
LOQUAT; JAPANESE MEDLAR—Native to China and Japan. Tree to 20'; leaves evergreen, handsome, to 10" long, rough, with brownish fuzz on underside, tooth-edged; flowers fragrant, small, white, in woolly spikes; fruit (February-March) up to 2" long, oval with slight neck, light- or dark-yellow, slightly downy; flesh usually white, very juicy, deliciously subacid; seeds, 1 or several, brown, glossy, to ⅝" long. Fruit eaten raw, cooked or made into preserves. Fast-growing from seed; superior types grafted. Fairly common as an ornamental fruit tree throughout Florida. For the past few years most of the fruit has been ruined by the Caribbean fruit fly.

Ervatamia coronaria Stapf. (*E.*   APOCYNACEAE
   divaricata Linn.; *Tabernaemontana coronaria* Willd.)
CRAPE JASMINE; EAST INDIAN ROSEBAY;
NERO'S CROWN—Considered probably native to India. Shrub, to 10' with milky sap; leaves evergreen, oval
or lance-shaped, pointed, up to 6 or 7" long, shiny;
flowers white, up to 2" across, ruffle-edged, clustered,
fragrant, especially at night. Double-flowered variety
flore-pleno preferred. Fruit an orange-colored pod; the
red pulp attached to seeds yields a red dye. Wood burned
for incense, also used medicinally, as are roots and
leaves; roots considered poisonous. Grown from cuttings.

Crape Jasmine
Ervatamia coronaria

Erythrina variegata var. *orientalis* Merr. (*E. indica* Lam.)   LEGUMINOSAE
CORAL TREE; LENTEN TREE; TIGER'S CLAW—Native to India and Malaya.
Tree, to 60', thorny; leaves deciduous, pinnate with 3 heart-shaped leaflets up to 6"
long; flowers red, 2 to 3" long with protruding stamens; flowers (early spring) in conspicuous, compact clusters to 1' long; fruit a black pod up to 1' long, twisted; seeds
red, used in novelties; poisonous if eaten. Leaves, bark, roots, medicinal. Grows fast
from seed, large cuttings, or air-layers. Branches often lumpy with insect galls.

Eucharis grandiflora Planch. & Lindl.   AMARYLLIDACEAE
AMAZON LILY—Native to Colombia. Herb, perennial, with nearly round, long-
necked bulb; leaves dark-green, long-stemmed, broad-oval, with pointed tip and indented longitudinal veins; flower stalk to 2' high, topped by cluster of exquisite white,
fragrant, drooping flowers with prominent stamen-bearing central cup surrounded
by 6 segments flaring to 3" across. Needs moist, shady location.

Eugenia axillaris Willd.   MYRTACEAE
WHITE STOPPER; WATTLE—Native to South Florida, the Bahamas and West
Indies. Tree, to 25' with gray bark, slender branches, densely foliaged; leaves
evergreen, pointed, to 3" long, dark-green, shiny, aromatic; new growth red; fruit
borne at base of leaves, black, ½" across, round, with minute crown at tip, juicy,
sweet, usually one-seeded. Fairly fast-growing from seed; salt-tolerant.

Eugenia foetida Pers. (*E. myrtoides* Poir.; *E. buxifolia* Willd.)   MYRTACEAE
BOX-LEAF EUGENIA, or SPANISH STOPPER—Native to southern Florida and
the Keys, the Bahamas, West Indies, Central America and Yucatan. Tree or shrub, to
20' or even 35', erect, compact; leaves evergreen, aromatic, opposite, oval, blunt, to
1¾" long; flowers tiny, white, in stalkless clusters; fruit nearly round, black, ¼"
wide, with 1 or 2 seeds. Easily grown from seed; very desirable for tall hedges and as a
tub plant in patios.

[73]

Eugenia paniculata Banks (*E. hookeri* Steud.)   MYRTACEAE
BRUSH CHERRY—Native to Australia. Tree, may exceed 40', of slender proportions; leaves evergreen, narrow, to 3" long, dark-green, new growth tinged with red; flowers to 1" across, in clusters of 3 to 5, with white, prominent stamens; fruit reddish-purple, to ¾" wide, used for jelly. Fast-growing from seed and cuttings. Can be clipped as a hedge.

Eugenia rhombea Urb.   MYRTACEAE
SPICEBERRY; RED STOPPER—Native to South Florida, the Bahamas and West Indies. Tree, to 25'; leaves evergreen, to 2½" long, dotted; flowers to ⅜" across, white, clustered; fruit orange-red, turning black, nearly round, up to ½" across. Wood used for furniture.

Eugenia uniflora Linn.   MYRTACEAE
SURINAM CHERRY—Native to Brazil. Shrub, or small tree, to 20' and equally broad, dense, bushy; leaves evergreen, highly aromatic, to 2" long, oval, pointed, glossy, new growth red; flowers solitary, small, white with prominent stamens; fruit (spring) rounded and flattened, ribbed, to 1¼" across, bright-red or very dark-red, shiny, thin-skinned, with juicy reddish-orange pulp, spicy, tart, sometimes sweet, excellent fresh or preserved. One round seed or two hemispherical. Commonly grown from seed throughout much of Florida, primarily as a hedge. Can be close-clipped and continues to fruit.

Euphorbia cotinifolia Linn.   EUPHORBIACEAE
RED SPURGE—Native from southern Mexico to northern South America. Shrub or tree, to 30', with abundant milky sap; leaves deciduous, opposite, more wedge-shaped than oval, to 5" long, ranging in color from purplish-green to maroon or blood-red on the upper surface and from pale to downy-white on the underside; flowers small, bell-shaped, ivory-white, in terminal clusters; fruit angled, hairy, containing purgative seeds. Entire plant, and especially the sap, apt to produce rash and blisters. Propagated by cuttings and, unfortunately, sold as an ornamental (since 1960) because of its colorful foliage.

Candelabra Cactus—*Euphorbia lactea*

Euphorbia lactea Haw. EUPHORBIACEAE
FALSE CACTUS; CANDELABRA CACTUS; MILKSTRIPE
EUPHORBIA—Native to East Indies. Succulent, cactus-like plant, to 15' with straight trunk and candelabra-like arrangement of 3- to 4-angled branches up to 3" wide, saw-edged and spiny, often with marbled white or yellow stripe on each surface. Rarely flowers (see photo); occasionally bears tiny, short-lived leaves. Exudes milky, sticky, toxic sap when cut. Variety *cristata,* the COCKSCOMB CACTUS or CRESTED MILKSTRIPE EUPHORBIA, has shortened, "crested", branches. Formerly believed to be caused by hail or other accidental or deliberate injury, the cresting (cristation) is apparently a hereditary trait reproducible from cuttings from vertical branches.

Euphorbia milii Ch. des Moul. (*E. splendens* Hook.) EUPHORBIACEAE
CROWN-OF-THORNS—Native to Madagascar. Shrub, to 4', with slender, twining, very thorny branches; leaves slender, paddle-shaped, up to 2" long; flowers (all year) minute, clustered, flanked by a pair of red, petal-like bracts, the whole seeming like a flower ½" wide. This plant sometimes distinguished as *E. splendens prostrata.* There are other erect forms with bracts that are salmon in color or yellow edged with red. A similar shrub with a flat, rectangular formation of flowers, with pink bracts has been identified as *E. keysii.* Grown from cuttings, in full sun with very little moisture.

Euphorbia pulcherrima Willd. (*Poinsettia pulcherrima* EUPHORBIACEAE
  Graham)
POINSETTIA; CHRISTMAS FLOWER—Native to Central America and northern Mexico. Shrub, to 12'; leaves more or less toothed or lobed, sometimes like traditional oakleaf, up to 7" long; the true flowers are small and greenish, or orange, in the center of a broad whorl of usually bright-red, slim, pointed bracts which latter are commonly thought of as composing the flower. There are double and triple forms which have bracts in dense peony-like clusters. Pink and ivory-white varieties are sometimes seen. Stems yield white, sticky juice which may cause dermatitis and is toxic internally, as are the leaves and bracts. Grown from cuttings; needs drastic pruning in late summer. Popular in South Florida, being very showy during the Christmas season and continuing so for many weeks. Universally grown as a pot plant.

Euphorbia tirucalli Linn. EUPHORBIACEAE
PENCIL TREE; MALABAR TREE; MILK BUSH; AFRICAN SPURGE
TREE—Native to Africa. Tree, to 30', becoming a great mass of green, fleshy, rubbery branches and branchlets, cylindrical, slim, curved out and upward, thornless; sometimes with a few small leaves. Milky sap is irritating to skin and will cause intense inflammation in the eyes. Grown from cuttings and unwisely planted in dooryards; becomes too large and is hazardous to prune.

Exostema caribaeum Roem. & Schult.   RUBIACEAE
PRINCEWOOD; CARIBBEE BARK TREE; JAMAICA BARK—Native to South
Florida, Bahamas and the West Indies, Central and northern South America. Shrub
or tree to 25'; leaves evergreen, oval, pointed, to 3" long and 1 ½" wide, may be slightly hairy beneath; flowers fragrant, white or pink-tinged, tubular, each slender tube, to
2" in length, splitting at its apex into 5 narrow lobes, 1" long, which curl backward
away from the 5 protruding stamens; seed pod ⅝" long, dark-brown. Wood strong,
hard, used for cabinetwork; bark bitter, has been used as substitute for cinchona in
treatment of fevers.

Ficus altissima Blume   MORACEAE
LOFTY FIG; FALSE BANYAN—Native to India, China and Philippines. Tree, to
80', with milky sap, wide-spreading branches; underground roots spreading and
destructive; aerial roots moderately produced, develop from mere hanging cords to
thick, trunk-like props; leaves evergreen, broad-oval, up to 8" long, leathery, glossy;
smooth, not downy, beneath; fruit small, round, bright-red. Fast-growing from cuttings and air-layers. Common street tree in Greater Miami.

Ficus aurea Nutt.   MORACEAE
FLORIDA STRANGLER FIG; GOLDEN WILD FIG—Native to South Florida,
the Bahamas and West Indies. Tree, to 65', stalwart, with milky sap and spreading
branches; aerial roots descend from its lower branches as it ages; leaves semi-
deciduous, oval, to 4" long, dark-green, shining; fruit up to ⅝" across, round,
orange-yellow or red, stemless, edible. Seed may sprout in any foothold on a palmetto
or other tree and, as the fig grows, its long, main roots, descending to the ground, em-
brace and finally smother its victim.

Ficus benghalensis Linn.   MORACEAE
BANYAN TREE: TRUE BANYAN—Native to India. Tree, to 85' with white, gummy sap, far-reaching, horizontal branches from which aerial roots descend to the
ground and thicken to form innumerable props for the massive top; leaves evergreen,
broad-oval, to 10" long, dark-green, shining above and, when mature, downy
beneath; fruit ½" wide, round, red, or rarely yellow, borne in pairs. Latex yields inferior rubber; bark and aerial roots yield fiber for cordage; latex, leaves and bark
medicinal. Propagated by cuttings and air-layers. Not as common in Florida as *F.
altissima*.

Strangler Fig
Ficus aurea

Benjamin Fig—*Ficus benjamina*

Ficus benjamina Linn.   MORACEAE
BENJAMIN FIG; WEEPING FIG—Native to India and Malaya. Tree, to 80', with milky sap and aggressive surface roots; forms a broad, dense dome of foliage on gracefully drooping, supple branches from which descend numerous aerial roots which become stout props after anchoring in the ground. Leaves evergreen, pointed, up to 5" long, glossy; fruit 1/3" wide, round, red, borne in pairs. May be close-planted and trimmed as a massive, soft hedge. Some particularly decorative forms of this species are erroneously known in the nursery trade as *"Ficus exotica"*. Propagated by air-layering or lopping off and rooting large branches. Fast-growing. Unwisely planted in small dooryards and requires frequent topping; as a street tree demands heavy maintenance to keep it under control, including root-pruning to avoid damage to pavement and driveways; root-pruned trees topple over in hurricanes. Fallen fruits are a nuisance and hazard on sidewalks.

Ficus carica Linn.   MORACEAE
FIG—Native to Asia Minor. Tree, to 30' with milky sap; leaves deciduous, broad, deeply 3- to 5-lobed, dull-green above, slight whitish down on underside which is netted with white veins; fruit borne close to branches, pear-shaped, up to 3" long, yellowish-green, reddish-brown, purplish or nearly black, soft, fleshy, sweet and mild-flavored, with tiny seeds in center. This is the common fig of which there are many commercial varieties. Grown only for home use in Florida. Propagated by cuttings; needs to be drastically pruned in winter.

India Rubber Tree
Ficus elastica decora

Shortleaf Fig
Ficus citrifolia

Fiddleleaf Fig
Ficus lyrata

Ficus elastica Roxb.   MORACEAE
INDIA RUBBER TREE—Native to tropical Asia. Tree, to 100', with milky sap,
stout trunk, long, spreading branches and aerial roots; main roots protrude above
ground and extend far from tree; leaves evergreen, thick, broad-oval with pointed
ends, to 12" long, dark-green, shiny, with numerous horizontal veins; young leaves
wrapped in slender, reddish, tube-like sheath before unfolding; fruit yellowish, up to
½" long, in pairs. Latex yields rubber of inferior grade. Propagated by cuttings and
air-layers. Formerly much grown as a potted house plant. Variegated-leaf varieties as
well as the non-variegated have been occasionally planted in South Florida in the
past. *F. elastica decora,* a type with broad leaves having ivory midrib which is red
beneath, and conspicuously red sheath on new growth, was introduced from Belgium
by Jim Vosters in 1952 and has been propagated by air-layering on a large scale for
dooryard landscaping and shipment as a house plant. It is often misused in planters
and patios where it quickly exceeds its space.

Ficus citrifolia Mill. (*F. laevigata* Vahl.; *F. brevifolia* Nutt.)   MORACEAE
SHORTLEAF FIG—Native to South Florida, the Bahamas and West Indies. Tree,
to 50', with milky sap; leaves semi-deciduous, oval, pointed at tip, to 4" long; fruit,
with short stem, round, up to ½" across, yellow at first, red when ripe, sweet and
edible. Fast-growing from seed; becomes a stately, broad-topped shade tree with few
or no aerial roots; should be appreciated and planted more commonly instead of
troublesome exotic species.

Ficus lyrata Warb. (*F. pandurata* Sander)   MORACEAE
FIDDLELEAF FIG—Native to tropical Africa. Tree, to 40', with milky sap, upright, open branches; leaves evergreen, clustered, leathery, fiddle-shaped, up to 1 ½'
long, with conspicuous whitish veins; fruit round, 1 ½ to 2" across, green with indented white dots, borne in two's. Propagated by cuttings and air-layers. This is a
highly ornamental, non-aggressive *Ficus* tree, satisfactory in patios and small
dooryards except for the continuous dropping of the huge old leaves. Often potted for
indoor decoration.

Ficus macrophylla Desf.   MORACEAE
MORETON BAY FIG—Native to Australia. Tree, growing to large size, with
slender upright trunk and roots spreading above ground; leaves evergreen, oval,
tapering at ends, up to 10" long and 4" wide, dark, shining, brownish on underside,
wrapped in reddish sheath before unfolding; fruit purple, dotted with white, roundish, up to 1" across, few in a cluster. Fast-growing from cuttings and air-layers.
Serves well as a pot plant with little watering.

Ficus pumila Linn. (*F. repens* Hort.)   MORACEAE
CREEPING FIG—Native to Japan, China, Australia. Climbing plant, with close-
clinging stems bearing leaves 1" long, and with short, outstanding branches with
leaves up to 4"; fruit pear-shaped or oblong with flattened end, up to 2 ½" long and
1 ½" wide; immature fruit broken open smells like coconut. Grows rapidly from cuttings. Attractive covering for stone walls; requires pruning.

Ficus racemosa Linn. (*F. glomerata* Roxb.)   MORACEAE
CLUSTER FIG—native to India and Burma. Tree, to 80', stalwart, with milky sap, very thick trunk and branches; leaves evergreen, not leathery, up to 7" long, broad at base, tapering to pointed tip; conspicuous veins; leafstems brown-fuzzy; fruit pear-shaped to near-round, up to 2" across, produced in dense clusters on trunk and branches, red when ripe, edible but not desirable.

Ficus religiosa Linn.   MORACEAE
PEEPUL; BO TREE; SACRED FIG—Native to India. Tree, to 100', erect, with smooth, light grayish-brown bark and compact head; few, if any, aerial roots; leaves deciduous, up to 6" long, somewhat heart-shaped with prolonged, very slender, "drip tip"; fruit purple, ½" across, borne in pairs. Fruit and leafbuds edible; bark used in tanning and dyeing and paper has been made from its fiber; latex used as sealing wax; various parts of tree medicinal. Fast-growing from leafbud or root cuttings and air-layers. Desirable as a street tree but not for home yards; shed leaves are quickly replaced by new growth which is light yellowish-green.

Ficus retusa Linn. (*F. nitida* Thunb.)   MORACEAE
INDIA LAUREL FIG—Native to India. Tree, to 100', forms a massive, densely foliaged and rounded head; has few aerial roots; leaves evergreen, oval, up to 4" long, shining; fruit 1/3" wide, yellow or dark-red, borne in pairs. Leaves, bark and aerial roots medicinal. Grows rapidly from cuttings and air-layers. A magnificent ornamental tree where adequate space is available, as in parks and school grounds.

Flacourtia indica Merr. (*F. ramontchi* L'Her.)   FLACOURTIACEAE
RAMONTCHI; GOVERNOR'S PLUM—Native to tropical Asia and Madagascar. Shrub or bushy tree, to 25' and equally broad, sometimes thorny, slender-branched; leaves evergreen, to 3" long, very shiny, the new growth tinged with red; flowers small, yellowish, inconspicuous; fruit round, to 1½" across, smooth-skinned, deep-red or very dark maroon; flesh light-brown, sweet or subacid; skin sometimes astringent; up to 10 seeds, flat, about ¼" long, buff, hard. Fruit eaten raw, cooked or preserved. Commonly grown as a tall "barrier" or untrimmed hedge. Bird-planted seedlings are invading South Florida hammocks.

Fortunella japonica Swingle   RUTACEAE
MARUMI KUMQUAT; ROUND KUMQUAT—Native to southern China. Shrub, dwarf, bushy, sometimes spiny; leaves evergreen, to 4" long and 1½" wide; flowers white, 5-petaled; fruit round, up to 1¼" across, brilliant orange-yellow, with acid flavor. Rind edible but pungent. Not to be confused with MEIWA or SWEET KUMQUAT which is also round but with a sweet rind. Propagated by grafting and air-layering. Often grown as an ornamental; fruiting sprays used to decorate gift baskets of *Citrus* fruits.

Fortunella margarita Swingle   RUTACEAE
NAGAMI KUMQUAT; OVAL KUMQUAT—Native to eastern Asia. Shrub or tree, to 12'; leaves evergreen, up to 6" long; flowers white, ½" across, single or a few

Coral Tree—*Erythrina variegata* var. *orientalis*

Governor's Plum—*Flacourtia indica*

Jacaranda—*Jacaranda mimosaefolia*

Lignum Vitae—*Guaiacum sanctum*

in a cluster; fruit cylindrical or oval, up to 1 ¾ " long, orange-yellow, with thick peel, scant acid pulp. Rind edible but sharply pungent. Makes excellent marmalade; sometimes preserved whole in brandy. Usually grafted on *Poncirus trifoliata* Raf., a much-used *Citrus* rootstock.

Galphimia glauca Cav. (*Thryallis glauca* Kuntze)   MALPIGHIACEAE
SHOWER-OF-GOLD—native to Mexico and Central America. Shrub, to 9', with slim branches, light-gray when mature, red-hairy when young; leaves evergreen, narrow-oval, thin, up to 2" long, on reddish stems; flowers (all year) yellow, 5-petaled, ¾ " wide, in showy clusters; fruit small, 3-lobed, splits open. Seedlings slow-growing; popular in foundation plantings. Not to be confused with GOLDEN SHOWER, *Cassia fistula* Linn., q.v.

Gardenia jasminoides Ellis   RUBIACEAE
CAPE JASMINE—Native to China. Shrub, to 6'; leaves evergreen, lance-shaped or oval, pointed, up to 6" long, dark-green, leathery, shiny; flowers (spring) up to 4" wide, somewhat like a double rose, with waxy, white petals, richly perfumed; fruit 1 ½ " long, ridged, orange, seldom produced. This is the popular "gardenia" of the florist trade; usually grafted on *G. thunbergia* in South Florida, otherwise needs hole especially prepared with acid soil. Requires special "gardenia" fertilizer.

Gliricidia sepium Steud. (*G. maculata* HBK)   LEGUMINOSAE
MADRE DE CACAO; QUICK STICK—Native to tropical America. Tree, to 25' with supple, drooping branches, heavily foliaged; leaves deciduous, pinnate with 3-11 oval leaflets up to 8" long, glossy green above, pale beneath and with pronounced coumarin odor. After leaves are shed in the dry season, the bare branches put forth an abundance of pink flowers, up to ¾ " across, in showy clusters. Fruit a flat pod up to 4" long. Fast-growing from seeds or large cuttings, the tree is much used as a shade for cacao and coffee and as a "living fencepost" in tropical countries. Flowers are edible; foliage medicinal; leaves and seeds used for poisoning rodents; wood valued.

Madre de Cacao—*Gliricidia sepium*

Gloriosa rothschildiana O'Brien   LILIACEAE
ROTHSCHILD GLORYLILY—Native to Uganda. Herb, tuberous-rooted,
slender-stemmed, climbing to 8 ft. and dying back in winter; leaves opposite or alternate, oblong, terminating in a curling tendril; flowers (spring-summer) gaudy, to 8"
wide, the 6 ribbon-like, recurving segments brilliant yellow-and-red, becoming totally
dull-red with age, slightly wavy; seed pods gained by hand pollination. Thrives in full
sun or partial shade. Tubers creep underground, extending and multiplying, and send
up plants some distance from original location. Tubers raised commercially in Central Florida. The MALABAR GLORYLILY (*G. superba* Linn.) is similar but has
crisped and more wavy segments; blooms in autumn. Plants and tubers are highly
toxic internally.

Graptophyllum pictum Griff. (*G. hortense* Nees)   ACANTHACEAE
CARICATURE PLANT; CAFE CON LECHE—Native to New Guinea. Shrub, to
10'; leaves oval, pointed,to 6" long, green variegated with yellow or white along the
midrib; a less common form has maroon leaves variegated with pink; flowers tubular, 1 ½" long, purplish-red, in compact clusters; fruit an oblong seed pod but rarely
produced. Grown from cuttings in partial shade.

Grevillea robusta A. Cunn.   PROTEACEAE
SILK OAK—Native to Australia. Tree, to 150', erect, with relatively short branches
and fern-like, evergreen foliage, the new growth silvery beneath; flowers bright-
yellow or orange in spikes to 4" long; fruit up to ¾" long. Wood valued for its
beautiful grain and luster. Slow-growing from seeds; not as common in South Florida
as further north in the State.

Guaiacum sanctum Linn.   ZYGOPHYLLACEAE
ROUGHBARK LIGNUM VITAE; HOLYWOOD—Native to South Florida, the
Bahamas and West Indies. Tree, to 30', with dense, umbrella top; bark grayish-white;
leaves evergreen, pinnate with leaflets up to 1" long, dark, glossy; flowers blue, 5-
petaled, ¾" wide, single or few in a cluster; fruit orange-yellow, ¾" long, 5-angled,
splits open revealing black seeds encased in brilliant-red arils. Wood highly valued for
its strength and durability; distinguished by areas of dark-brown in contrast with
light honey-color. Very slow-growing, from de-ariled seeds. Rare in cultivation.
Blooms and fruits more or less continuously throughout the year.

Gynura aurantiaca DC.   COMPOSITAE
VELVET PLANT—Native to Java. Herb, perennial, to 3', its stems covered wtih
lavender or purple fuzz; leaves to 6" long, toothed, soft, velvety, the hairy surface
having a purple tinge; flower-heads ¾" wide, yellow or orange, on ends of stems rising to a foot above the foliage. Fast-growing from stem cuttings or leafbuds in full
sun, with little water. An old-time favorite in South Florida gardens.

Hamelia patens Jacq. (*H. erecta* Jacq.)   RUBIACEAE
SCARLET-BUSH; FIREBUSH—Native to South Florida and down to northern
South America. Shrub, to 12'; leaves evergreen, oval or oblong, pointed, to 8" long,

generally in whorls of 3, hairy, often purplish or red-tinted; flowers (all year) bright reddish-orange, tubular with slight flare, up to 1½" long, in clusters to 5" across; fruit round-oval, ¼" long, dark-red or purple, almost black, seedy, edible. Grown from seed, cuttings or air-layers.

Harpullia arborea Radlk. and *H. pendula* Planch.　　　SAPINDACEAE
TULIPWOOD—Native to Asia and East Indies, and to Australia, respectively. Tree, to 60', with dense crown of drooping branches; leaves evergreen, alternate, compound with 5 to 9 oblong leaflets, short-pointed at tip, glossy, to 6" long; flowers tiny, greenish, inconspicuous; fruit red-orange, 2-lobed, to 1" wide, hollow except for 2 black seeds. Fine street and dooryard trees mainly emanating from specimen on Montgomery estate, Old Cutler Road, which came from a nursery in California (as *H. arborea*) more than 30 years ago; and from seeds obtained by Edwin Menninger from New South Wales (as *H. pendula*).

Hedychium coronarium Koenig　　　ZINGIBERACEAE
GINGER LILY—Native to India. Herb, with fleshy rhizome and erect, succulent stems to 6'; leaves alternate, strap-like, pointed, to 2' long and 4" wide; flowers in terminal spikes, white, very fragrant, with very slender tube 3" long, 3 narrow outer segments and 3 broad inner segments flaring to a width of 4"; seed capsule oblong, to 2", containing bright-red seeds. Propagated by division or seed; forms dense clumps in sun or shade; has become naturalized in moist areas.

Heliconia spp.　　　HELICONIACEAE
(formerly MUSACEAE)
FALSE BIRD-OF-PARADISE—Native to tropical America. Herbs, to 15' or more, with fleshy, erect stems in clumps; leaves oblong, 10" to 4' long, entirely green or variegated, red-veined, or red or purple on the underside; inflorescence erect or hanging, consisting of orange, red or variously colored, boat-shaped bracts in 2 overlapping rows, from which protrude green, pink, red, orange or yellow flowers. Several species grown in South Florida, especially in moist locations. *H. latispatha* Benth., with erect inflorescence, is most common and least spectacular; *H. rostrata* Ruiz & Pavon, with hanging inflorescence, is very showy. Propagated by division.

Hemerocallis spp.　　　LILIACEAE
DAY LILY—Native, mostly, to China and Japan. Herbs, perennial, with fibrous or fleshy tubers and tufts of grass-like leaves (deciduous or evergreen), 1 to 3' long; flower stalks, to 2' tall, bear, in spring, a succession of funnel-shaped, 6-parted flowers to 4" long—in various shades of yellow, orange, pink, red or purple; some variegated. May be propagated by offsets which develop on old flower stalks or by root division. Should receive sun in the morning; shade in the afternoon.

Hibiscus cannabinus Linn.　　　MALVACEAE
KENAF—Native to tropics of Old World. Shrub, usually annual, to 14', spiny; upper leaves deeply 5-lobed, lower leaves heart-shaped; flowers yellow or red with bright-red center; fruit round, bristly, up to ¾" long. Plant widely cultivated for jute-

like fiber; of recent interest as a source of paper, and now grown in South Florida farmlands for its stems which are used as bean poles replacing *Sesbania emerus.* Leaves edible; seeds used in sauces and yield an oil.

Hibiscus eetveldeanus Wildem. & Th. Dur.   MALVACEAE
   (*H. acetosella* Welw.)
RED-LEAF HIBISCUS—Native to South Africa. Shrub, to 8', with upright, un-branched, smooth, magenta-red stems; leaves somewhat maple-like, 5-lobed, tooth-edged, to 3" long, also magenta-red; flowers 3" wide with 5 overlapping petals slight-ly more lavender-red than the rest of the plant and with a dark "eye"; open in mor-ning and close at noon. After flowers fall, a conical, ¾"-long, hairy seed pod develops within the red, ribbed calyx; calyx bears a basal fringe of slim, outstanding, forked bracts. This plant, commonly grown from seed for its color in South Florida gardens, is often confused with Roselle (*H. sabdariffa,* q.v.). The first published record of the species in the Western Hemisphere appeared in the first edition of this book.

Hibiscus elatus Sw. (*Paritium elatum* Don)   MALVACEAE
TREE HIBISCUS; BLUE MAHOE; MOUNTAIN MAHOE—Native to highlands of Cuba and Jamaica. Tree, to 35'; leaves evergreen, heart-shaped, with prolonged tip and round-toothed edge, up to 8" across and 10" long; flowers 4" long, orange when first open, turning blood-red and then dark-maroon by the end of the day; fruit a hairy seed pod ¾" long. Timber valued for its long grain, flexibility and color; heartwood is olive-green variously tinged with blue or purple; prized for furniture, salad bowls and implements, gunstocks, ladders, etc.; fiber from bark called "Cuba bast"; shoots and foliage have medicinal uses. Fast-growing from seeds and cuttings; rare in South Florida.

Hibiscus mutabilis Linn.   MALVACEAE
COTTON ROSE; CONFEDERATE ROSE—Native to China. Shrub, to 15'; leaves deciduous, irregularly heart-shaped, to 8" wide, velvety, brownish beneath; flowers (late fall) to 4" wide, white or pink in morning, turning dark-red by nightfall, may be single or double; fruit a round, hairy pod, 1" across. Bark yields fiber. Grown from seeds or cuttings; should be pruned back after blooming. Rare in South Florida.

Hibiscus pilosus Fauc. & Rend. (*H. spiralis* Cav.)   MALVACEAE
Native to Florida Keys, West Indies and Mexico. Shrub, to 6' tall, hairy; leaves heart-shaped or 3-lobed, tooth-edged, up to 1½" long; flowers bright-red, pendent, to 1" long, like slightly unfurled rosebuds, and with protruding staminal tube; seed pod ⅜" long.

Hibiscus rosa-sinensis Linn.   MALVACEAE
CHINESE HIBISCUS; SHOE-FLOWER—Native to Asia. Shrub, to 30'; leaves evergreen, broad-oval, tapering to pointed tip, tooth-edged, sometimes 3-lobed, up to 5" long; flowers (all year) borne singly, to 6" or more in width, with large, overlap-

ping, delicate petals and silky throat from which a slender and usually red tube extends about 4", with yellow stamens and red pistil arranged at its tip. Petals may be white, yellow, pink, salmon, red or variegated and the throat pink, dark-red or white. Some varieties have double, ruffled blossoms. Most last for a single day, closing at night. Flowers pickled and eaten in China and food-coloring is made from the red flowers and lime juice; petals yield a shoe-blacking and mascara; leaves, roots, bark and flowers medicinal. One of the most commonly cultivated ornamental and hedge plants in Florida. Propagated by seeds, cuttings, air-layers and grafting. There are numerous cultivars and hybrids between this species and *H. schizopetalus*. *H. rosasinensis* var. *cooperi* Nichols has attractive foliage, variegated with white and red, and pendent, rich-red flowers. During the past 10 years, many individual bushes and hedges have been lost due to the hand-grenade scale, *Cerococcus deklii,* which attacks the nodes and causes a steady decline.

Hibiscus sabdariffa Linn.   MALVACEAE
ROSELLE; JAMAICA SORREL—Native to tropics of Old World. Annual plant, to 8', with upright red stems; leaves green, deeply 3-lobed, up to 3" long; flowers yellow, followed by thickening of the red calyx which becomes fleshy (November) and is used for cranberry-like sauce, juice, jelly, and for wine. The seed pod which it encloses is best removed before cooking. Seeds may be roasted and eaten; also yield a healing oil. Leaves edible and have medicinal uses; stems yield fiber. Formerly so common in Florida home gardens as to be nicknamed "Florida Cranberry". Rarely grown today. Raised from seeds sown in late May or June.

Hibiscus schizopetalus Hook. f.   MALVACEAE
FRINGED HIBISCUS—Native to East Africa. Shrub, to 12', with slim, drooping branches; leaves evergreen, slender to broadly oval, pointed, toothed; flowers, to 4" wide, hang from long, slender stems; petals pink or red, deeply cut, almost fringed, curl backward, away from the long slim tube which bears the stamens and pistil. Grown from cuttings.

Hibiscus tiliaceus Linn. (*Paritium tiliaceum* Juss.)   MALVACEAE
MAHOE; SEA HIBISCUS—Native to South Florida, the Keys and other tropical shores of both hemispheres. Shrub, or spreading tree, to 50', of coastal habitat; bushy, with drooping branches that bend to the ground and take root in low, wet land, thus repeating, further inshore, the land-building action of the mangroves. Leaves evergreen, rounded heart-shaped with short tip, smooth-edged, to 6" wide, light on underside; flowers (all year) up to 4" wide, yellow, with or without maroon eye, turning dark-rose when they fall at the end of the day; seed pod velvety, to ¾" long, distributed by ocean currents. Leaves and root medicinal; leaves edible, when young, and also used for fodder; bark yields fiber widely used for cordage; wood light in weight and used for floats, also in boats and cabinetwork. Fast-growing from seeds or cuttings; on dry ground becomes a handsome, round-topped shade tree, if lower branches are pruned (see photo on page 86).

Mahoe Tree—*Hibiscus tiliaceus*

Hippomane mancinella Linn.   EUPHORBIACEAE

MANCHINEEL—Native to South Florida, the Bahamas, West Indies and Central America. Tree, to 40' with compact head; leaves deciduous, alternate, oval, to 4" long; flowers tiny, yellowish or rose, in 6" spikes; fruit nearly round, up to 1 ½" across, green or yellow-green, sometimes blushed with red; thin-fleshed with woody core enclosing a 6- to 8-seeded, spiny stone. Tree and fruit have milky sap that is internally and externally poisonous. Rare, having been largely exterminated as a dangerous tree.

Holmskioldia sanguinea Retz   VERBENACEAE

MANDARIN HAT, or CHINESE HAT PLANT—Native to the subtropical Himalayan region. Shrub, to 15' with long, arching branches which may be trained to a trellis; leaves nearly evergreen, opposite, somewhat downy, ovate, square at the base, pointed at the tip, fine-toothed, to 4 ½" long; flowers in terminal clusters (mainly in winter), red, tubular, 1" long, rising from the center of a saucer-like calyx ¾" across. Seeds generally absent in Florida. Grown from cuttings, ground-layers or air-layers.

Hoya carnosa R. Br.   ASCLEPIADACEAE
WAX PLANT—Native to Australia and China. Vine, climbing by aerial roots, to
15'; leaves evergreen, opposite, fleshy, smooth, oval, pointed, 2 to 4" long; flowers,
fragrant, waxy, star-shaped, ½" wide, white with red center, borne in flat, drooping
clusters. Petiole cuttings and leaves root quickly; vine slow-growing; succeeds in any
soil but is highly susceptible to nematodes. Needs partial shade and little water.
Flowers rich in nectar, attract humming-birds.

Hura crepitans Linn.   EUPHORBIACEAE
SANDBOX TREE—Native to tropical America. Tree,
to 100', with clear, sticky sap and thick trunk covered
with sharp projections; branches wide-spreading; leaves
semi-deciduous, heart-shaped with pointed tip, to 6"
long; flowers maroon; female flower fleshy, resembles
small toadstool indented in center and with toothed,
recurved edge; male flowers in 2"-long cone, like
miniature ear of corn; fruit round, flattened, up to 3"
across with ribbed, woody, brown shell; explodes noisily
when mature and scatters the seeds. Sap toxic, causes
dermatitis, used to stupefy fish; seeds cause severe inter-
nal poisoning. Immature fruit in early days was used to
hold sand for drying ink, in lieu of blotters. Occasionally
and unwisely cultivated in dooryards, from seed.

Hylocereus undatus Britt. & Rose   CACTACEAE
NIGHT-BLOOMING CEREUS; STRAWBERRY PEAR—Considered as possibly
native to Mexico. Cactus, with crawling or climbing, triangular, jointed stems about
3" wide, scalloped, with sharp thorns on the edges, and producing aerial roots;
flowers (late summer) spectacular, up to 1' across, white, with many petals, a fluff of
long stamens, and a prominent style with a rosette of stigmas at its tip; bloom opens
at night and perfumes the air; fruit oval, to 4" long, with deep-pink rind composed of
overlapping scales; pulp white, juicy, filled with tiny black seeds. Fruit and flower
bud edible; juice of plant medicinal. Grows readily from cuttings.

Night-blooming Cereus—*Hylocereus undatus*

Hymenocallis latifolia Roem. (*H. keyensis* Small)   AMARYLLIDACEAE
KEYS HYMENOCALLIS; KEYS   SPIDER-LILY—Native to southeastern
Florida and the Keys. Herb, perennial, with round bulb and strap-like, arching leaves
up to 2½' long; flower stalk erect, topped by a cluster of a dozen or more flowers;
flowers white with slender tube to 6", and narrow, long, downward-curving sepals
and petals, and erect filaments 2½" long connected at their bases by a delicate web.
Propagated by bulb division.

Ilex cassine Linn.   AQUIFOLIACEAE
DAHOON HOLLY—Native to swamps and stream banks from southeastern
Virginia to the Florida Keys, the Bahamas and Cuba, also around the Gulf Coast to
Louisiana. Tree, to 40', erect, compact; leaves evergreen, alternate, leathery, smooth,
oblong or elliptic, to 5" long; flowers, small, white, clustered among the leaves, male
and female usually on separate plants; fruit (in fall) round, to ⅜" wide, generally red,
abundant and showy. The only red-berried native holly in South Florida; successfully
grown on high ground if given adequate water. Grown from cuttings, or seedlings
may be grafted. Fine for tall hedges, stands close-clipping.

Bush Morning-Glory
Ipomoea crassicaulis

Horsfall Morning-Glory
Ipomoea horsfalliae

Ipomoea crassicaulis Robinson CONVOLVULACEAE
BUSH MORNING-GLORY—Native from Brazil and
Peru to Mexico. Shrub, with slender stems to 8'; leaves
semi-deciduous, elongated heart-shaped, to 6" long;
flowers (all year) bell-like, up to 3" wide, fragile, pink
with purplish throat. Fast-growing from cuttings. A
similar shrub, *I. leptophylla* Torr., native to the western
U. S., develops a massive tuber which enables it to sur-
vive in arid regions.

Ipomoea horsfalliae Hook.   CONVOLVULACEAE
HORSFALL MORNING-GLORY; PRINCE'S
VINE—Native to tropical America and possibly also to
tropics of Old World. Vine, with huge tuberous root and
slender, climbing and twining stems; leaves evergreen,
divided into 5 to 7 lance-shaped, pointed, wavy-edged
leaflets, spread like the open fingers of a hand, up to 6"
long, dark-green, glossy on top, light-green beneath;
flowers trumpet-shaped, to 2½" long, in clusters, rose or
purple, silky. Variety *briggsii* has very attractive, fuchsia-
colored flowers. Seeds rarely produced. Propagated by
cuttings or air-layers.

Ipomoea pes-caprae Roth   CONVOLVULACEAE
RAILROAD VINE; BEACH MORNING-GLORY;
GOAT'S-FOOT MORNING-GLORY; BAY
HOPS—Native to Georgia, Florida, Texas, the
Bahamas, West Indies, tropical America and Old World

tropics. Herb, succulent, creeping, sometimes extending rope-like stems 60 or 70' along a sandy beach; can be trained over a fence or trellis; leaves evergreen, rounded, notched at tip, suggesting a goat's foot, up to 4" wide, held aloft on 3- to 4"-stems; flowers bell-shaped, purple or dull-pink, up to 2" long; fruit a round-oval capsule about ½" long. Seeds medicinal; plant and roots edible but considered toxic in quantity. Grown from seed; a useful sandbinder and salt-tolerant vine for coastal locations.

Railroad Vine
Ipomoea pes-caprae

Ipomoea tuberosa Linn.   CONVOLVULACEAE
YELLOW MORNING-GLORY; WOOD ROSE—Native to the tropics generally. Vine, with leaves to 8" wide, deeply cut into as many as 7 pointed segments; flowers (in fall) bell-shaped, up to 2" long; yellow; dried brown sepals with rounded brown seed pod in center suggest rose carved from wood; much used in dried arrangements. Grown from seed in full sun, quickly covers large area.

Iresine herbstii Hook.   AMARANTHACEAE
CHICKEN GIZZARDS—Native to Brazil. Subshrub, annual, to 6' high; leaves opposite, slightly fleshy, with prominent, curved veins, nearly circular, notched at tip, to 2½" long, purple-red with light-red veins and red stems, or entirely bright-yellow; flowers minute in large sprays, but usually absent. Grown from cuttings. Showy plant for borders or hedges in full sun; can be sheared to any height. Commonly cultivated in Latin American gardens.

Iresine lindenii Van Houtte   AMARANTHACEAE
BLOODLEAF—Native to Colombia and Ecuador. Subshrub or herb, annual, compact; usually with deep-red stems; leaves lance-shaped, sharp-pointed, with prominent veins, generally deep-red, rarely yellow with light-green areas and red stems. Grown from cuttings. Effective in close-sheared beds in full sun, especially in parks and parkways.

Ixora coccinea Linn.   RUBIACEAE
RED IXORA; JUNGLEFLAME IXORA; BURNING LOVE—Native to southern Asia. Shrub, to 15' with small branches, compact growth; leaves evergreen, narrow, to 4" long; flowers (all year) scarlet, 4-petaled, up to 1½" across, in compact, flat clusters about 4" broad; fruit round, purple-black, ½" wide, edible. Grown from seed, root suckers or cuttings as an ornamental and hedge plant, often pruned in columnar form. Prefers partial shade but blooms more profusely in sun; often shows iron deficiency. Variety *lutea* has yellow flowers and supple, drooping branches.

Ixora duffyi T. Moore (*I. macrothyrsa* auth. NOT T. Moore)   RUBIACEAE
MALAY IXORA—Native to East Indies. Shrub, to 10' with spreading branches; leaves evergreen, oblong or lance-shaped, pointed, up to 1' long, glossy; flowers (all year) flaring to 1" wide at end of slender 2"-tube, deep-red, numerous, in flat clusters to 8" across. Fruit red. Grown from seeds, cuttings or air-layers; needs acid soil.

Jacaranda mimosaefolia D. Don (*J. acutifolia*   BIGNONIACEAE
   Humb. & Bonpl.
SHARPLEAF JACARANDA—Native to Brazil. Tree, to 50' with slim trunk and
limbs, spreading top; foliage fine, feathery, shed during the winter; flowers (mainly
spring; also in August) lavender-blue, slender bell-shaped, up to 2" long, abundant,
in large, conspicuous sprays; fruit a brown, circular pod up to 2" across which splits
open. Grown from seed or cuttings and sometimes grafted. A prized ornamental
which seems less adapted to southeastern Florida than to other sections of the State.
Seed pod halves are used for decorative purposes.

Jacobinia mexicana Seem.   ACANTHACEAE
Native to Mexico. Shrub, with leaves to 6" long, oval or lance-shaped, with long,
pointed continuation extending down the leafstem; flowers tubular, slender, up to
1½" long, bright-red. Grown from cuttings.

Jacquinia barbasco Mez (*J. armillaris* Jacq.)   THEOPHRASTACEAE
BARBASCO—Native to seacoasts of the West Indies and northern South America.
Shrub or tree to 15' with light-gray bark; leaves gray-green, oval, thick, leathery, to 4"
long; flowers fragrant, white or yellowish, ¼" wide, in terminal sprays; fruits (fall
and winter) scarlet, in showy, drooping clusters, round, to ½" wide. Slow-growing
from seed. Introduced from St. Vincent by Dr. David Fairchild and Harold Loomis
in 1930 and, during the past 20 years, has been advocated for beach planting because
of its salt-tolerance and attractive fruits. All parts of tree toxic and much used in the
past as fish poisons.

Jacquinia keyensis Mez   THEOPHRASTACEAE
JOEWOOD; CUDJOE-WOOD—Native to South Florida and Bahamas. Shrub or
tree, to 15', with light-gray bark; leaves evergreen, narrow, paddle-shaped, to 4" long,
stiff; flowers ½" across, ivory, fragrant, in spikes; fruit round, ⅜" wide, scarlet or
yellow.

Jasminum fluminense Vell (*J. bahiense* DC.;   OLEACEAE
   erroneously *J. azoricum*)
Native to Brazil. Vine, with slender, wiry stems, climbing to treetops; leaves
evergreen, opposite, compound, with 3 ovate, pointed leaflets to 2" long, often hairy
beneath; flowers (all year) in loose clusters, white, very fragrant, with slender tube
1¼" long, the 8 segments flaring to a width of 1"; fruit round, black, abundant. Fast-
growing; may be propagated by cuttings, but in Florida and Puerto Rico it self-sows
its seed, has spread extensively and become a major pest.

Jasminum multiflorum Andr. (*J. pubescens* Willd.)   OLEACEAE
DOWNY JASMINE—Native to India. Shrub, with downy, sprawling or climbing
stems; leaves evergreen, opposite, pointed-oval, to 3" long; flowers (all year) scarcely
perfumed, white, 1" wide, with 4 to 9 lobes, usually borne in clusters. Grown from
cuttings; may be close-pruned and still bloom profusely. The similar *J. gracillimum*
(illustrated), from North Borneo, blooms during the winter.

[90]

Downy Jasmine—*Jasminum gracillimum*   Arabian Jasmine—*Jasminum sambac*

Jasminum officinale Linn. OLEACEAE
COMMON JASMINE; POET'S JASMINE—Native from western Asia to China. Shrub, with supple, drooping or semi-climbing branches; leaves deciduous, compound with 5 to 7 pointed leaflets up to 2½" long; flowers (all year) white, 1" wide, star-shaped on slender tube, borne in open clusters, heavily fragrant. Easily grown from cuttings.

Jasminum officinale var. *grandiflorum* Bailey OLEACEAE
  (*J. grandiflorum* Linn.)
CATALONIAN JASMINE; ROYAL, ITALIAN or SPANISH JASMINE—Native to southern Asia. Shrub, fairly upright, with supple branches; leaves deciduous, pinnate, with slender-pointed leaflets; flowers star-shaped, to 1½" across, white, purple-tinted, fragrant, used in making perfume. Grown from cuttings.

Jasminum sambac Ait. OLEACEAE
ARABIAN JASMINE—Native to India. Climbing shrub to 5'; leaves evergreen, glossy, oval, to 3" long; flowers up to 1" wide, few or several in a cluster, white at first, changing to purple, fragrant. Flowers yield perfume and are used by the Chinese in tea; leaves, roots and flowers used medicinally. The Grand Duke jasmine is a double-flowered variety resembling a compact, white rose. Grown from cuttings.

Jasminum volubile Jacq. (*J. gracile* Andr.; OLEACEAE
  erroneously *J. simplicifolium*)
WAX JASMINE—Native to Africa. Shrub, to 5', bushy, compact, or climbing; leaves evergreen, in pairs, oval, slightly pointed, smooth, shining, to 1½" long; flowers fragrant (all year), white, to ¾" long, the lobes narrow and far apart; fruits (rare) small, round, black. Popular as a closely-trimmed low hedge or border or in planting bins because of its fine foliage; flowers are not showy.

[91]

Jatropha curcas Linn.   EUPHORBIACEAE
PHYSIC NUT; BARBADOS NUT—Native to Bermuda, West Indies, tropical
America, and Old World tropics. Tree, to 15', with sticky, yellowish sap, that turns
red after exposure to air; leaves deciduous, broad, with 3 to 5 lobes; flowers small,
greenish-yellow; fruit (summer) oval, to 1½" long, yellow, turning black as it dries
and splits open revealing 2 or 3 oblong seeds about ¾" long, of good flavor but
dangerous to eat because of poisonous and purgative properties; less potent if
roasted. Oil from seeds used for illuminating, also in paints and soaps; leaves used to
stupefy fish, also as medicinal poultices. Grown from seed or cuttings and unwisely
planted in dooryards.

Jatropha gossypifolia Linn. (*Adenoropium*   EUPHORBIACEAE
    gossypifolium Linn.)
BELLYACHE BUSH—Native to West Indies and tropical America. Herbaceous
shrub, to 5'; leaves deciduous, to 6" wide, 3- to 5-lobed, green, glossy; new growth
dark-purple in one form; flowers small, deep-red or purple, in small clusters; fruit ⅜"
in diameter, blunt-oval, with 6 lengthwise ridges and 3 small seeds dangerous to eat
because of toxic and purgative effect. Sap, leaves, stems and roots used in folk
medicine.

Jatropha integerrima Jacq. (*J. hastata* Griseb.;   EUPHORBIACEAE
    J. pandurifolia Andr.)
PEREGRINA—Native to Cuba. Shrub or small tree, to 15', with slender trunk and
branches; leaves evergreen, fiddle-shaped or irregularly lobed, up to 6" long; flowers
(all year) 1" across, in upright clusters, bright-red; fruit round-oval, 6-lobed, up to
¾" long, containing smooth, speckled toxic seeds which are scattered as the dry fruit
explodes. Grown from cuttings in full sun.

Jatropha multifida Linn.   EUPHORBIACEAE
CORAL PLANT—Native from southwestern U.S. to northern South America.
Shrub, to 15' high and nearly as broad; leaves semi-deciduous, up to 1' across, deeply
and finely cut; flowers (spring-fall) small, bright coral-red in erect, open, coral-
stemmed clusters; fruit tri-cornered, about 1½" across, green when unripe, turning
yellow before it falls, contains 1 to 3 round, light-brown seeds which are dangerously
purgative and toxic. Grown from seed or cuttings and commonly planted in
dooryards.

Jatropha podagrica Hook.   EUPHORBIACEAE
GOUT PLANT—Native to Central America. Shrub, to 4', the woody stem swollen
at the base; leaves few, deciduous, 3- to 5-lobed, to 10" wide, pale beneath, with
leafstem, to 1½' long, attached off-center; flowers small, orange-red, in erect, flat
heads; fruit oval, 1" long, containing smooth, toxic seeds. Grown from seed as a
curiosity, in full sun; drought-tolerant.

Juniperus chinensis var. *japonica* Lav.   CUPRESSACEAE
JAPANESE JUNIPER—Native to Japan. Shrub, dwarf, to 1' high, with reclining
branches spreading to 3 or 4'; the shoots densely set with evergreen leaves, some scale-

like, some needle-like in 3's. Much grown from cuttings as a ground cover, in sun or shade. There are several named cultivars, the most popular being known as Pfitzer's Creeping Juniper. Not as successful in South Florida as further north.

Juniperus conferta Parl. (*J. litoralis* Maxim.)   CUPRESSACEAE
SHORE JUNIPER—Native to coasts of Japan. Shrub, with reclining stems and short branches rising to 18"; branchlets densely clothed with needle-like, spiny-tipped leaves, bluish- or silvery-green; fruit round, to ½" wide, black with a powdery bloom, 3-seeded. Grown from cuttings as a ground cover, especially desirable for sandy soil, thrives in full sun; defoliates in shade; needs plenty of water and also spraying to control red spider; can be kept low and compact by clipping.

Juniperus silicicola Bailey   CUPRESSACEAE
SOUTHERN RED CEDAR—Native to coasts from North Carolina to central Florida and west to Mississippi; also eastern Texas. Tree, to 50', pyramidal; leaves evergreen, aromatic, tiny, pointed and silvery on new shoots, scale-like and dark-green on older twigs; cone nearly round, fleshy, dark-blue with lighter bloom, favored by birds. Grown from seed. Excellent as ornamental tree and clipped as a hedge. Salt-tolerant. Yields useful wood and cedar oil. Copious pollen gathered by honeybees.

Southern Red Cedar—*Juniperus silicicola*

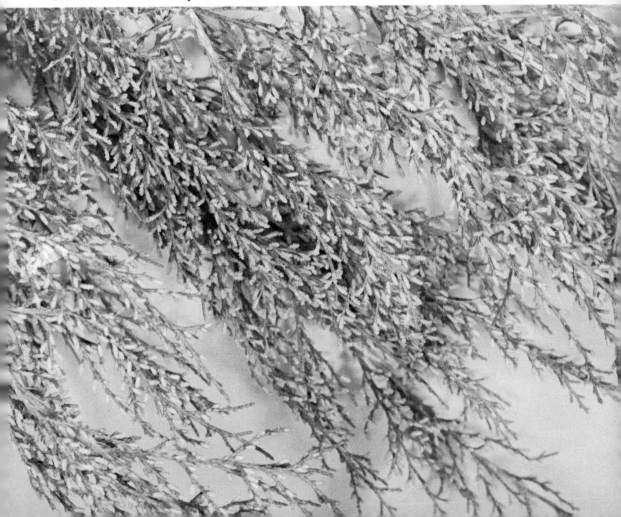

Kalanchoe beharensis Drake del
    Castillo   CRASSULACEAE
Native to Madagascar. Shrub or small tree, to 20'; the stem roughened by projecting, spiny leaf scars; leaves, to 1' or more, triangular, partly folded, trowel-like, with lobed and wavy edges, thick green, covered with silvery felt underneath, rusty wool above. Rarely blooms. Needs full sun. Propagated by leaf cuttings or offshoots; or, if plant becomes too tall and leggy, may be beheaded and the top rooted to make another plant.

Kalanchoe fedtschenkoi Hamet & Perr.   CRASSULACEAE
Native to Madagascar. Herb, perennial, succulent, to 1½'; stems flexible, partly reclining and spreading; entire plant coated with grayish or bluish bloom; leaves fleshy, nearly circular, bluntly toothed; flowers salmon, tubular, in drooping clusters. Stems and leaves root easily. Common ground cover or rock garden plant; useful in sunny, dry situations.

Kalanchoe pinnata Pers. (*Bryophyllum pinnatum* Kurz)
                              CRASSULACEAE
LIVE-FOREVER; LIFE-PLANT; CATHEDRAL BELLS—Native to tropics of Old and New World. Succulent plant, to 6' high; leaves pinnate with leaflets up to 5" long, fleshy, triangular, with scalloped edge; erect flower stalk topped with cluster of flowers like slender, hanging bells, the calyx tube-like, 1½" long, light-green flushed with purplish-red; protruding from it are the wine-red petals, slightly flaring at the tip. Plant spreads like a weed; new plants will sprout from a plucked leaf, and the leaves are accordingly often sold as curiosities to pin on a curtain. Juice of plant used medicinally.

Kigelia pinnata DC.   BIGNONIACEAE
SAUSAGE TREE; FETISH TREE—Native to tropical Africa. Tree, to 50', forms broad head; leaves evergreen, pinnate with blunt leaflets to 6" long; flowers night-blooming with heavy, grassy scent, funnel-shaped, up to 4" wide, brownish- or purplish-red, suspended on very long stems; hand-pollination results in abundance of inedible, woody fruits which resemble sausages and may be 2' or more in length and 5" in width; surface light grayish-brown. Fruit used medicinally when young. Fast-growing from seed, the tree attracts much attention as a curiosity in South Florida.

Koelreutaria elegans A. C. Smith (*K. formosana* Hayata)   SAPINDACEAE
GOLDEN RAIN TREE—Native to Formosa and Japan. Tree, to 60'; leaves deciduous, to 2' long, twice compound, with toothed leaflets 1½ to 3" long; flowers (October-November) small, yellow, in large, erect sprays; fruit 3-lobed, papery, balloon-like, becoming pink or quite red, longer-lasting and more showy than the blooms, and contains 3 round, black seeds. Fast-growing from seed. An admirable dooryard tree, increasingly planted in the past 15 years.

Sausage Tree—*Kigelia pinnata*

Lagerstroemia indica Linn.   LYTHRACEAE
CRAPE MYRTLE—Native to southern Asia and northern Australia. Shrub, tree-like when old and in some areas may reach 35'; leaves deciduous, narrow, up to 3" long; flowers (summer) to 1½" across, with 6 flat-spreading, fringed petals, form showy, conical sprays up to 9" in length, abundant; color ranges from white or lavender to different shades of pink or near-red; fruit a roundish, woody seed-capsule, ½" across. Grown from seeds and cuttings in full sun; will not flower if pruned. An old-time favorite in Florida and other southern states.

Lagerstroemia speciosa Pers. (*L. flos-reginae* Retz.)   LYTHRACEAE
QUEEN'S CRAPE MYRTLE; QUEEN-OF-THE-FLOWERS—Native to India and East Indies. Tree, to 60' or more with broad, rounded top; leaves broad-oval or lance-shaped, blunt, to 1' long; turn vivid red and fall after cold spells; flowers 6- or 7-petaled, crinkly, up to 3" wide, pink or lavender, rarely white, abundant in large, pyramidal, erect clusters; fruit round, to 1" across, turns black and splits open to release seeds. Tree lovely in full bloom (June-July), also yields valuable wood; roots, seeds, bark and leaves medicinal. Slow-growing from seed, cuttings or root shoots. Rare in Florida.

White Mangrove—*Laguncularia racemosa*

Laguncularia racemosa Gaertn.f.   COMBRETACEAE
WHITE BUTTONWOOD; WHITE MANGROVE—Native to coasts of Florida,
the Bahamas, West Indies, Central and South America and Mexico. Tree, to 30', with
gnarled trunk, brown bark; leaves evergreen, oval, rubbery, up to 3" long; flowers
tiny, in spikes, greenish, velvety, fragrant; fruit greenish-brown, velvety, ridged, ¾"
long; bark contains tannin; wood strong and close-grained.

Lantana camara Linn. (including *L. aculeata* Linn.)   VERBENACEAE
COMMON LANTANA—Native from southern U.S. to northern South America
and widespread throughout the tropics. Shrub, to 6' or much taller in the wild state;
usually prickly stemmed; leaves evergreen, to 5" long, oval, tapering to a point, fine-
toothed, rough, pungently aromatic; flowers on upright stems in small, flat, nosegay
clusters, cream, yellow or pink, turning to orange, scarlet or purple, the clusters being
two-toned as the color-change progresses from the outer flowers to those in the
center; fruit round, dark-blue, to ¼" wide, toxic when green, edible when ripe. The
entire plant is toxic to cattle, sensitizing their skin to sun injury and causing liver
damage. Occurs as a weed in Florida and is unwisely grown as an ornamental.

Lantana depressa Small   VERBENACEAE
DWARF LANTANA—Native to South Florida pinelands and Keys. Subshrub, 8"
to 3' tall, with hairy, creeping stems often forming low, dense, mats; leaves aromatic,
opposite, to 1 ½" long, oval or elliptic, pointed, blunt-toothed, finely hairy above and
below; flowers (all year) small, rich-yellow, in flat clusters less than 1" across; fruit
round, small, not abundant. Grown from seed or cuttings as a ground cover.

Lantana involucrata Linn.   VERBENACEAE
WILD SAGE; BIG SAGE—Native from southern U.S. to northern South America
and West Indies. Shrub, to 5 ½'; with rough, downy, toothed leaves up to 1 ¼" long,
pungently aromatic; flowers lavender or white in small clusters; fruit 3/16" wide,
dark-blue.

[96]

Fringed Hibiscus
Hibiscus schizopetalus

Mandarin Hat
Holmskioldia sanguinea

Red Ixora
Ixora coccinea

Tulipwood—*Harpullia pendula*

Crape Myrtle—*Lagerstroemia indica*

Sabicu—*Lysiloma sabicu*

Texas Silverleaf—*Leucophyllum frutescens*

Kanapalei
Mimusops roxburghiana

Panama Berry
Muntingia calabura

Queen's Wreath
Petrea volubilis

Lawsonia inermis Linn.   LYTHRACEAE

HENNA; EGYPTIAN PRIVET—Native to Africa and
Asia. Shrub, to 25', sometimes spiny; young stems
square; leaves evergreen, narrow, pointed, up to 1½"
long; flowers ½" across or less, with 4 crinkly petals,
yellow, white, rose or orange-red, in clusters up to 1'
long, delightfully fragrant. Leaves and stems yield henna
used for dyeing hair, fabrics, etc.; flowers yield perfume;
various parts of plant used in folk medicine. Grown from
seeds or cuttings; pruning encourages flower production;
may be trimmed as a hedge.

Leucaena leucocephala DeWit (*L. glauca* Benth.)   LEGUMINOSAE

LEAD TREE; JUMBIE BEAN—Native to West Indies, tropical America and
naturalized in the Bahamas and South Florida. Tree, usually to 30', may attain 65';
foliage feathery; flowers white, in fluffy balls up to 1½" across; fruit a narrow,
reddish-brown pod up to 6" long containing flat, brown seeds. Young pods and seeds
and new shoots cooked as vegetables; plant grown as fodder for cattle and goats but
causes hair loss in horses, mules, donkeys and pigs; seeds used in bracelets and other
novelty wear.

Leucophyllum frutescens I. M. Johnston   SCROPHULARIACEAE
  (*L. texanum* Benth.)

TEXAS SILVERLEAF, or CENIZA—Native to southwestern Texas and adjacent
areas of Mexico. Shrub, to 9', with downy-white stems; leaves slender, to 1" long,
silvery; flowers (mainly in winter), bell-shaped, mauve, to 1" wide. Slow-growing
from cuttings; thrives in dry, sandy soil; needs full sun, occasional pruning. Rare in
1950; available in several nurseries 10 years later; and is frequently planted today.

Ligustrum japonicum Thunb.   OLEACEAE

JAPANESE PRIVET—Native to Japan. Shrub, to 10', bushy; leaves evergreen,
round or oblong to 4" long, leathery, waxy, dark on top, lighter below; flowers white,

Japanese Privet—*Ligustrum japonicum*

small, in loose clusters up to 6" long; fruit small, deep-blue, rarely develops in South Florida. Commonly grown from cuttings as a hedge and foundation shrub. Since 1960, increasingly used as a small tree, trimmed to form. This species has been mistakenly sold as *L. lucidum* Ait., a hardier, very fruitful species grown further north.

Ligustrum sinense Lour. (erroneously distributed as   OLEACEAE
    L. amurense Carr.)
CHINESE PRIVET—Native to China and Korea. Shrub, to 15', with slender, spreading downy branches; leaves semi-evergreen, oval or oblong, blunt, downy on underside of midrib when young, to 3" long, dull grayish-green; flowers small, white, in downy, loose clusters to 4" long; fruit oval 1/3" long, with slight bloom. Propagated by cuttings or air-layers. Occasionally grown as a hedge in South Florida; more commonly further north. Stands close pruning.

Litchi chinensis Sonn.   SAPINDACEAE
LYCHEE, or LITCHI—Native to southern China. Tree, to 35', with dense, rounded crown; leaves evergreen, compound, with 2 to 4 pairs of elliptic to lance-shaped, pointed leaflets to 5" long; flowers greenish-ivory in terminal sprays to 1' long; fruit (June-July) broad-oval, with minutely knobby, leathery skin, bright-red when ripe, becomes brown and brittle when dry; flesh fragrant, grapelike, white-translucent, luscious, subacid; drying to color and consistency of raisin; highly prized fresh, canned or dried. First propagated by air-layers in Florida by Col. W. R. Grove in 1940; has since become a much-appreciated fruit tree for dooryards and commercial plantings.

Livistona chinensis R. Br.   PALMAE
CHINESE FAN PALM—Native to China. Palm tree; trunk to 30' high and 1' or more in diameter, bearing old leaf bases or, later, merely ringed; head compact, rounded; leaves fan-like, circular, up to 6' across, finely divided half-way to center, and with stout stems which have short recurved spines near the base until old; the fringe-like ends of the leaves droop, giving the palm its typical shaggy aspect; flowers small, greenish, in long clusters; fruit dark turquoise, oval, 1" long. Seeds viable 4-6 weeks. Often grown as a tub plant in the North.

Japanese Honeysuckle—*Lonicera japonica*

Lonicera japonica Thunb.   CAPRIFOLIACEAE

JAPANESE HONEYSUCKLE—Native to eastern Asia, naturalized in the southeastern United States. Shrub, trailing or climbing to 30 ft., stems hollow, twining, hairy when young; leaves semi-evergreen to evergreen, opposite, downy on both sides when young, on underside only when mature, ovate or oblong, blunt or pointed, to 3 ½" long; flowers very fragrant, borne in pairs in leaf axils near the ends of the branchlets, white or pink at first, turning yellow with age, tubular, 2-lipped, to 1 ½" long; fruit black, shining, round, 3/16" wide. Fast, rampant grower from seeds or cuttings, in sun or shade, blooms from spring to fall; drought- and wind-tolerant; can be grown on fence or trellis or used as a ground cover or soil-retainer on steep road banks.

Lysiloma latisiliquum Benth. (*L. bahamensis* Benth.)   LEGUMINOSAE

WILD TAMARIND—Native to South Florida, the Bahamas and West Indies. Tree to 60', erect, with slender trunk; bark gray at first, later brown, scaly; branches speading and drooping gracefully; foliage feathery, deciduous, twice-pinnate with leaflets ¼ to ½" long, pale on the underside; flowers tiny, white, in fuzzy balls about ½" across; fruit a thin, flat pod, to 5" long, reddish-brown, containing several glossy brown seeds ½" long. Grown from seed; salt-tolerant; rare in cultivation but should be more commonly planted.

Lysiloma sabicu Benth. (*L. latisiliqua* Benth.)   LEGUMINOSAE

SABICU—Native to Cuba, the Bahamas, Hispaniola, Puerto Rico, Trinidad and Yucatan. Tree, to 20' (much taller in Cuban forests), with slender trunk and wide crown of gracefully weeping branches; leaves semi-deciduous, twice-pinnate, feathery, with oblong leaflets to ¾" long; young growth brownish to rose-red and showy; flowers small, white, in globose heads; seed pod flat, thin, to 4" long, containing brown seeds. When borne in abundance, the dry pods rustle in the wind but less noisily than those of *Albizia lebbeck*. Slow-growing from seed; salt-, wind- and drought-tolerant; has been planted along highways and in parks; deserves wider use in landscaping. Wood has been much used for boat-building in Cuba.

Wild Tamarind—*Lysiloma latisiliquum*

Macadamia integrifolia Maiden & Betche   PROTEACEAE
   (SMOOTH MACADAMIA NUT)
Macadamia tetraphylla L. Johnson
   (ROUGH MACADAMIA NUT)
Native to Queensland and New South Wales. Tree, to 30'; leaves evergreen, leathery, oblong, to 12" long, more or less spiny-toothed; flowers (early spring) small, in fuzzy, cream-white, dangling spikes; fruit in elongated clusters, with thick green husk splitting open and releasing light-brown, round, hard-shelled nut to ¾" wide, containing crisp, delicious, oily kernel. Grown experimentally by private individuals and horticultural stations. Most Florida trees set scanty crops. Propagated by grafting. Grown commercially in Hawaii.

Magnolia grandiflora Linn.   MAGNOLIACEAE
                                    (formerly LAURACEAE)
SOUTHERN MAGNOLIA—Tree, to 90', of stately, pyramidal form; leaves evergreen, alternate, leathery, dark-green, glossy, oval, pointed, to 8" long; flowers fragrant, white, waxy, with 6 to 12 broad petals to 4" long; fruit oval, cone-like, to 4" long, containing red, ½"-long seeds. Occasionally grown in South Florida. Seedlings take several years to bloom but air-layers and grafts flower early and this beautiful tree should be seen more often in our area. Prefers acid soil but one specimen transplanted from Melbourne in mid-1930's continues to flourish in limestone at Palm Lodge Tropical Grove, Homestead.

Southern Magnolia—*Magnolia grandiflora*

Malpighia coccigera Linn.   MALPIGHIACEAE
SINGAPORE HOLLY—Native to the West Indies but widely grown in Malaya and
Java long before its adoption into South Florida horticulture. Shrub to 4', with
slender branches; leaves evergreen, leathery, glossy, spiny-toothed like miniature holly; flowers light-pink, to ¼" across; fruit round, red, ⅛" wide. Slow-growing from
seed or cuttings. Ideal for dwarf hedges and pot culture. Prostrate forms (offered
since 1954) are valued as ground covers, some adapted to shade, others to full sun.

Malpighia punicifolia Linn.   MALPIGHIACEAE
BARBADOS CHERRY; WEST INDIAN CHERRY—Native from southwestern
U.S. to northern South America and West Indies. Shrub or tree, to 12', bushy, with
slender branches and narrow, evergreen leaves to 3" long; flowers to ¾" wide with
delicate white or pink petals with scalloped edges; fruit cherry-like, 3-lobed, up to
1 ¼" across, bright-red to dark purplish-red, thin-skinned, with juicy pulp, tart, subacid, or nearly sweet; seeds in 3 triangular, ridged stones. Fruit outstandingly rich in
vitamin C, excellent for juice, jelly and puree for infants. Fairly common in South
Florida as a fruiting shrub for the home. Propagated by cuttings or air-layers.

Malvaviscus arboreus Cav.   MALVACEAE
TURK'S-CAP; SLEEPING HIBISCUS—Native to Mexico. Shrub, to 10' or more,
bushy, hibiscus-like; leaves evergreen, to 5" long, broad-oval with long-pointed tip,
toothed; flowers (all-year) bright-red, like partly open hibiscus blooms, to 2½" long,
pendent; fruit red, mucilaginous, edible, but not known in Florida. Grows readily
from cuttings; fairly common hedge plant in South Florida; more disease-resistant
than the Chinese hibiscus.

Malvaviscus drummondii T. & G. (*M. arboreus*   MALVACEAE
    var. *drummondii* Schery)
DRUMMOND WAX MALLOW—Native to southern U.S. including Florida; also
Mexico and West Indies. Herb, perennial, to 9', more or less downy; leaves somewhat
heart-shaped but 3-lobed, velvety; flowers orange-red, 1" long with protruding
staminal tube, resemble very small, partly open hibiscus blooms, pendent; fruit
round, red, edible, splits open.

Mangifera indica Linn.   ANACARDIACEAE
MANGO—Native to southern Asia. Tree, to 90' and attaining a spread of 100';
leaves evergreen, narrow, pointed, to 15" long, in dense clusters and recurved or
drooping; new leaves red or yellowish; flowers small, fragrant, ivory to reddish, in upright sprays which grace the tree with a frothy mantle when in full bloom (winter);
fruit (May to September) green, yellow, red, purplish or variegated, 2 to 9" long, with
juicy, peach-like flesh, in some varieties smooth, in others undesirably fibrous; flavor
varies from deliciously sweet or subacid to sour and ranges from mildly to markedly
"turpentiny"; stone large, flattened, more or less bearded. Propagated by seeds and
by grafting. Popular dooryard and commercial fruit in South Florida. Sap of tree,
skin and even juice of fruit irritating to the skin of some people.

Manihot esculenta Crantz   EUPHORBIACEAE
CASSAVA; TAPICOA PLANT; MANIOC; "YUCA" (Spanish)—Native to Brazil.
Herbaceous shrub, sometimes tree-like, to 10' or more; leaves up to 16" across, deeply cut into 3 to 7 lobes; flowers small, cupped, in elongated clusters; roots tuberous,
somewhat cylindrical, starchy, usually poisonous when raw but edible when properly
cooked or made into flour or tapioca. Bitter Cassava and Sweet Cassava (the latter
containing little or no prussic acid) were formerly distinguished as separate species.
Now grown commercially on a small scale in South Florida.

Manilkara bahamensis Lam. & Meeuse (*Mimusops emarginata*   SAPOTACEAE
     Britt.)
WILD DILLY—Native to South Florida, the Bahamas and West Indies. Shrub or
tree, to 30'; leaves evergreen, clustered at ends of twigs, slender-oval, notched at tip,
leathery, smooth on top with brownish fuzz on underside; flowers up to ¾" across,
yellow, clustered; fruit round, 1¼" wide, brown, scurfy of surface, with brownish
pulp and 1 to 4 flat seeds. Reportedly edible. Tree and immature fruit contain sticky
white latex.

Manilkara zapota Royen (*Achras zapota* Linn.)   SAPOTACEAE
SAPODILLA; NASEBERRY—Native to Mexico and Central America. Tree,
reaching 60', densely foliaged with evergreen, stiff, glossy, dark-green leaves up to 5"
long, pointed, and in rosette-clusters; flowers small, cream-colored; fruit (mainly in
summer) round, oval or oblate, 2 to 4" in diameter, light- or dark-brown with sandy
surface; pulp yellow-brown, greenish- or reddish-brown, juicy and very sweet; seeds
shiny, black, oblong, flat, often with small barb, should not be swallowed. Sticky
white latex of tree and unripe fruit yields chicle for chewing gum; wood valued for
construction; tree storm-resistant and handsome; cultivated for its fruit and as an ornamental. Superior selections propagated by grafting.

Mascarena verschaffeltii Bailey (*Hyophorbe verschaffeltii*   PALMAE
     Wendl.)
SPINDLE PALM; PIGNUT PALM—Native to Mauritius. Palm tree, to 30', with
smooth, light-gray trunk bulging toward the top and with a stout, usually pale-green
crownshaft extending another 2' or so to the base of the leaves; leaves feather-shaped,
up to 7' long and 4' wide, stiffly arched; flowers orange in a foot-long, brush-like
cluster; fruit black, oblong, 1¼" in length with cylindrical seed. Seeds viable for 2-3
months.

Melaleuca quinquenervia S. T. Blake (formerly known as *M.*   MYRTACEAE
     leucadendron Linn.)
CAJEPUT, or CAJUPUT; PUNK TREE—Native to Australia and southeast Asia.
Tree, to 80' or more, erect, with upturned branches; whitish, shaggy, thick bark;
leaves evergreen, slim, stiff, to 8" long, white-silky when young; spicily aromatic;
flowers (most profuse in October-November and slightly less so in June-July) small,
with prominent stamens, usually ivory-white, in bottle-brush clusters up to 6" long,
followed by spikes of small, hard, brown capsules containing minute seeds. Leaves

[102]

Cajeput—*Melaleuca quinquenervia*

yield cajeput oil, used medicinally; bark used for insulation; wood valued for cabinetwork. Fast-growing from seeds. Popular in South Florida landscaping and running wild in the Everglades. A common cause of respiratory trouble when in bloom.

Chinaberry—*Melia azedarach*

Melia azedarach Linn.   MELIACEAE
CHINABERRY; PRIDE-OF-INDIA; PERSIAN LILAC—Native to southern
Asia. Tree, to 60', forming dense, symmetrical, umbrella top; leaves deciduous, compound, with small, attractive, pointed, toothed leaflets; flowers lavender in loose clusters up to 6" long with lilac fragrance; fruit yellow, round, to ¾" across, abundant, in large clusters, conspicuous when tree is partly or entirely defoliated. Fruit poisonous; seeds used as beads; all parts of plant used medicinally. Fast-growing from seeds or cuttings. Very common dooryard tree in Key West, and frequently seen throughout Florida and other southern states.

[104]

Melicoccus bijugatus Jacq. (*Melicocca bijuga* Linn.)   SAPINDACEAE
MAMONCILLO; SPANISH LIME; GENIP—Native to tropical America. Tree, to
60', with rounded head; leaves evergreen, compound with 4 leaflets, pointed-oval, up
to 4" long; leafstems often flanged with oblong, leaf-like wings; flowers whitish, in
spikes (male and female on separate trees); fruit (summer) in large, grape-like
clusters, round or oval, ¾ to 1" across, with green, leathery skin and juicy, whitish or
pinkish, sweet or acid pulp; one large, round seed or 2 half-spheres, with edible
kernel. Conspicuous street tree in Key West; rare on the mainland.

Metopium toxiferum Krug.   ANACARDIACEAE
   & Urb.
POISONWOOD—Native to South Florida, the
Bahamas and West Indies. Tree, to 40', with reddish-
brown or gray, thin, flaky bark; leaves evergreen, pinnate, with 3 to 7 wedge-shaped leaflets up to 3½" long,
dark-green, glossy as if varnished; flowers (spring) tiny,
yellowish, in sprays; fruit oval, to ¾" long, dull-orange,
in loose, dangling clusters. This tree is often mistaken for
a young Gumbo Limbo. The resinous sap of all parts of
the tree is clear until exposed to the air when it turns
black. Old or injured leaves are dotted with blackened
sap. Contact may produce severe rash or blisters. Very
common in the pinelands and hammocks of the east
coast from Palm Beach south and in the Keys.

Mimusops roxburghiana Wight   SAPOTACEAE
   (*Manilkara roxburghiana* R. N. Parker)
KANAPALEI; RENGA—Native to India. Tree to 30'; leaves evergreen, oblong with
rounded tip, to 8" long, dark and glossy above, grayish beneath; leathery; fruit
round, smooth, yellow, to 1½" wide, with dry, mealy, yellow, sweet, edible flesh and
4 brown, triangular seeds ¾" long. Excellent wind-resistant, salt-tolerant tree for
coastal planting.

Mirabilis jalapa Linn.   NYCTAGINACEAE
FOUR-O'CLOCK; BUENAS TARDES ("good afternoon"); MARVEL-OF-
PERU—Native to tropical America. Herb, with turnip-like root and fleshy, upright
stems to 3', bushy; leaves somewhat triangular, up to 6" long; flowers white, yellow,
red or variegated, trumpet-like, 5-parted, flaring to 1" across at end of 1 to 2" tube;
open in late afternoon, fragrant, the aroma at night believed to repel mosquitoes;
fruit round, black. Leaves and roots medicinal; seeds poisonous; face powder made
from seed kernels in the Orient. Grown from seed as an ornamental plant and occurs
also as a weed in South Florida.

Monstera deliciosa Liebm.   ARACEAE
CERIMAN—Native to Guatemala and Mexico. Climbing or rambling plant with
stout stems bearing aerial roots; leaves deeply cut and perforated, up to 4' long and 3'

wide. What is frequently mistaken for the bloom is an erect, stiff, white bract, suggesting the spathe of a calla-lily and up to 14" or so in height, which hoods the floral spadix; the latter develops into a fruit a foot or more long and up to 3½" wide with a green, tiled rind, which is shed as the fruit ripens; the white, segmented flesh is fragrant, sweet and delicious, though throat-irritating when underripe due to its calcium oxalate content. The tough, central core is not eaten. This plant is more commonly cultivated as an ornamental than for its fruit in South Florida and is grown as a pot plant in the North; the leaf design is commonly used on textiles. Easily grown from seeds (not always present) and from cuttings.

Morinda citrifolia Linn.   RUBIACEAE
INDIAN MULBERRY; LIMBURGER TREE—Native to southern Asia, East Indies and Australia. Tree, to 20'; leaves broad, up to 10" long, glossy, soft and rippled; flowers white, star-shaped, arranged in compact head which develops into a compound fruit, oblong, up to 4" long and 2½" wide or more, bumpy of surface, with many "eyes", pale greenish-white when ripe, soft and odoriferous like old cheese but edible. Leaves and roots medicinal; bark yields red dye, roots, a yellow dye. The tree is tender to cold; grown as an oddity in gardens on the Florida Keys.

Moringa oleifera Lam.   MORINGACEAE
HORSERADISH TREE; BEN TREE—Native to India. Tree, to 25', of slender proportions; branches drooping; foliage feathery, evergreen; flowers 1" wide, white, in pendent clusters, fragrant; fruit edible when young and tender like stringbean, later becomes a dry, 3-sided, light-brown pod up to 18" or so in length and containing dark-brown, pea-sized, winged seeds. The tree grows readily from seed and cuttings, blooms and fruits continuously and abundantly. Flowers and foliage edible as "greens"; root is pungent like true horseradish; seeds yield ben oil, used for lubrication and culinary purposes, also in toilet products.

Indian Mulberry—*Morinda citrifolia*

Horseradish Tree—*Moringa oleifera*

Morus rubra Linn.   MORACEAE
RED MULBERRY—Native over large area of U.S., south from Massachusetts to
southern Florida and west to Texas. Tree, to 70' with slender, drooping branches;
leaves deciduous, heart-shaped, pointed, to 6" long, toothed, soft and silky; flowers
minute, in a small catkin; fruit (late spring) oblong, up to 2" long and ½" broad, red
when unripe, black when ripe, subacid, delicious, like a long blackberry without the
seediness. Fast-growing from cuttings.

Muntingia calabura Linn.   ELAEOCARPACEAE
PANAMA BERRY; CAPULIN; JAMAICA CHERRY—Native to tropical
America. Tree, to 30', soft-wooded, with slender, spreading, drooping branches;
leaves evergreen, narrow-oval, pointed, fine-toothed, soft, green on top, downy-white
on underside, up to 5" long; flowers white, resembling those of the strawberry, up to
¾" wide; fruit single or in pairs, round, to ⅝" across, red or pale-yellow, thin-
skinned; pulp sweet, juicy, filled with very minute seeds. Tree very fast-growing from
seed; bark yields fiber for cordage.

Murraya paniculata Jack (*M. exotica* Linn.)   RUTACEAE
ORANGE JESSAMINE; MARILLA—Native to southern Asia, Australia and
islands of Pacific. Shrub or tree, to 20' with slender proportions, densely foliaged;
leaves evergreen, pinnate, with pear-shaped leaflets up to 1½" long, dark-green,
glossy; flowers white, 5-petaled, small, clustered, very fragrant; fruit oval, to ½"
long, red, inedible. Grown from seeds or cuttings. Common in South Florida as an
ornamental shrub or small tree; excels all others as a hedge plant.

Musa nana Lour. (*M. cavendishii* Lamb.)   MUSACEAE
DWARF BANANA; CAVENDISH BANANA; CHINESE DWARF
BANANA—Native to southern China. Herb, to 7', stalk, composed of overlapping
leaf-bases, thick and stubby; leaves to 4' long; flowers yellowish-white in tiered rows,
each row shielded by a purplish bract, brownish-red on inside; an unopened, purplish
bud remains at tip of cluster; fruit up to 6" long and 1½" across, may be upwards of
200 in a bunch. The most commonly cultivated banana in Florida. Propagated by
corm division; should have protection from wind.

Musa paradisiaca Linn. MUSACEAE
PLANTAIN; COOKING BANANA—Native to India. Herb, to 30'; leaves up to 9'
long and 2' wide; flowers ivory, in tiered rows on thick stalk, each row shielded by
lavender bract which may be red on inner side; fruit to 14" in length and 2½" wide,
somewhat curved like a cow's horn and with a pointed tip; vivid green when immature, yellow when ripe; up to 80 in a bunch. Flesh firm, yellowish, subacid, edible
raw but usually cooked either unripe or ripe; also made into meal for invalids. Rare in
Florida.

Musa paradisiaca var. *sapientum* Kuntze MUSACEAE
BANANA—Native to India. Herb, to 20 or 30', depending on variety, with leaves to
12' long and 2' or more wide; fruit usually yellow, flecked with brown when ripe. The
Lady Finger variety is a tall, slender plant with small bunches and fruit only 3" in
length, thin-skinned and very sweet, rare in Florida; the Apple is very similar in plant
and fruit but the latter is plumper and distinguished by its pleasantly acid, apple-like
flavor, often grown for home use; the commercial Gros Michel is a stouter plant with
thicker-skinned fruit to 8" long and up to 200, even 300, in a bunch, seldom grown in
Florida; the Red Jamaica is dull-red of skin, somewhat shorter and plump, also rare;
the Orinoco, Horse, Burro or Hog banana is a tall but sturdy plant usually with a few-
handed bunch of thick, 3-angled fruits about 6" in length which are subacid, eaten
raw and, like the Plantain, good for cooking; hardy and rather common.

Myrica cerifera Linn. MYRICACEAE
SOUTHERN WAX MYRTLE, or SOUTHERN BAYBERRY—Native to coasts
from New Jersey to South Florida and Texas; also the Bahamas, West Indies and Bermuda. Shrub, or tree to 40'; if given room will form a globular clump 20' high and
wide; leaves evergreen, aromatic, alternate, narrow, to 4" long, irregularly toothed;
flowers tiny, in small spikes among the leaves; fruit (in winter) round, to ⅛" wide,
green, coated with bluish wax used for bayberry candles. Abundant in the wild; salt-
tolerant; may be grown from seed for coastal planting.

Nandina domestica Thunb. NANDINACEAE
HEAVENLY BAMBOO—Native to Japan, central China and India. Shrub, not a
bamboo, with erect stems in clumps, to 8' high, reddish-green when young; leaves
clustered at the tops of the stems, evergreen but red in winter, to 20" long and 30"
wide, twice- or thrice-compound; leaflets elliptic, leathery and to 3" long; fruits
(autumn), round, ⅜" wide, bright-red or yellow, abundant and showy. Slow-growing
from seed; stands drastic pruning. An old favorite in subtropical gardening, not sufficiently appreciated in South Florida.

Nephrolepis exaltata var. *bostoniensis* Davenport OLEANDRACEAE
(formerly POLYPODIACEAE)
BOSTON FERN—Origin unknown, may be a hybrid; *N. exaltata* is native to South
Florida and the tropics generally. Fern, of the sword-fern group, with erect but
arching, flat, coarse, feather-shaped fronds, to 3'tall and 8"wide, dotted on underside
with brown spore-clusters. Many cultivated varieties. Valuable as a ground cover in
shaded areas but multiplies rapidly by runners and requires control.

Oleander—*Nerium oleander*

Nerium oleander Linn.   APOCYNACEAE
OLEANDER—Native from southern Europe to western Asia and possibly eastward
to Japan. Shrub, to 20', branching almost vertically from near base; leaves evergreen,
to 8" long, slim, pointed, stiff, dark-green; flowers (spring-fall) to 3" wide, in clusters;
petals usually white, pink or rose, sometimes double; seed pod slender, to 6" long,
ridged lengthwise. All parts of the plant are internally poisonous to man and animals.
Grown from cuttings in full sun; salt-tolerant. A too-common ornamental in Florida.

Noronhia emarginata Stadtm.   OLEACEAE
MADAGASCAR OLIVE—Native to Madagascar.
Tree, to 30'; leaves evergreen, leathery, oval, notched at
tip, edges recurved, to 7" long; flowers small, yellowish,
in short clusters; fruit oval, flat at base, 1¼" long,
dark-purple when ripe; with cream-colored, edible pulp.
Slow-growing from seed; more wind- and salt-tolerant
than the seagrape; recommended for coastal locations.

Ochrosia elliptica Labill.   APOCYNACEAE
Native to Oceania. Tree, to 20'; leaves evergreen, oblong, blunt, leathery, glossy, to
6" long; flowers yellowish-white, ½" long, in flat clusters, fragrant; fruit, borne in
pairs, meeting end-to-end, oval, somewhat flattened, pointed, up to 2" long, brilliant-
red, smooth-skinned, with white, mealy flesh and large, tear-drop-shaped stone
which is polished as a pendant in Hawaii. Fruit reported to be somewhat toxic;
should not be confused with the Carissa (q.v.) which it outwardly somewhat
resembles. Grown from seed, usually as a pruned shrub; salt-tolerant.

Odontonema strictum Kuntze (erroneously *Jacobinia,*   ACANTHACEAE
     or *Justicia, coccinea*)
CARDINAL FLOWER—Native to Central America. Shrub, to 6', with upright
stems; leaves to 6" long, pointed at both ends, sometimes wavy; veins light, conspicuous and often slightly hairy on underside; flowers tubular, up to 1" long, bright-
red, clustered in a spike near the top of a slim, dark-purple flower stalk which extends
several inches above the leaves. Propagated by division or cuttings.

Snuffbox Tree
Oncoba spinosa

Oncoba spinosa Forsk.   FLACOURTIACEAE
SPINY ONCOBA; SNUFFBOX TREE—Native to
tropical Africa and western Asia. Shrub or tree, to 20',
branches drooping, thorny; leaves semi-evergreen,
pointed-oval, finely toothed, young leaves reddish;
flowers, 2 to 3" across, suggest fried egg, having large
center of yellow stamens surrounded by white petals;
fruit inedible, hard-shelled, reddish-brown when ripe,
round, up to 2" across, yellow and seedy within. Shell
used as snuffbox and for child's rattle. Roots and leaves
medicinal. Grows rapidly from seed; rare in South
Florida.

Ophiopogon japonicus Ker-Gawl.   LILIACEAE
MONDO GRASS, or DWARF LILYTURF—Native
from the Himalayas to Japan. Herb, with small, round
tubers; leaves evergreen, in tufts, grass-like, to 1' long,
dark-green; non-blooming in South Florida. Since 1956,
much planted as a ground cover, especially in places too
shady for grass. Propagated by division. Will not stand
foot traffic.

Pachira aquatica Aubl.   BOMBACACEAE
GUIANA CHESTNUT; MALABAR CHESTNUT; PROVISION TREE—Native
to tropical America. Tree, to about 30', stocky and spreading; leaves evergreen, compound, with 5 to 7 leaflets, to 12" long, spread like the fingers of a hand; flowers (opening in evening or early morning and closing at noon) fragrant, with 10"-long, strap-like, recurved petals and prominent tuft of long, white stamens tipped with red filaments; fruit a pod to 5" across and 15" long, compartmented, filled with pulp-covered seeds, ½" wide; seeds eaten raw or roasted. Rare in South Florida.

Pandanus baptistii Hort.   PANDANACEAE
BAPTIST SCREWPINE—Native to South Sea Islands. Herbaceous plant, with short stem hidden by a clump of upright, arching, 1"-wide, ribbon-like leaves, blue-green, striped with yellow or white, smooth-edged. Grown from suckers in full or partial sun with little watering.

Pandanus sanderi Mast.   PANDANACEAE
TIMOR SCREWPINE—Native to East Indies and Timor. Herbaceous plant, short-stemmed and densely tufted with ribbon-like leaves 2½' long with minute spines on edges; alternating green and yellow stripes run from leaf-base to tip. Often grows in clumps. Prefers full sun and dry soil.

Pandanus tectorius Soland.   PANDANACEAE
BEACH SCREWPINE—Native to seacoasts from southern China to tropical Australia and Polynesia. Tree, to 25' with spreading branches and numerous prop roots; forms large, dense thickets; leaves evergreen, spirally set, strap-like, to 5' long, spiny-edged, drooping; female flowers in a single spike with a yellow spathe; male inflorescence, on separate plant, highly fragrant, composed of many dangling spikes in long, white spathes; fruit, a globose, knobby head to 10" long, orange-yellow, breaking apart when fully ripe, exposing soft, edible pulp in center. Terminal bud edible. Leaves much used for thatching and mats. Grown from cuttings.

Pandanus utilis Bory   PANDANACEAE
COMMON SCREWPINE—Native to Madagascar. Tree, to 60', with stilt-like prop roots; branches few, sleek, rounded, tipped with clusters of evergreen, strap-like, spiny leaves, 3' long and 3" wide; fruit nearly round, up to 6" across, compound and rough-surfaced, green, yellowing as it ripens, with small amount of edible pulp, numerous large seeds. Leaves used for matting and baskets; roots yield fiber for cord and weaving. Slow-growing from seeds or large cuttings.

Common Screwpine—*Pandanus utilis*

Pandanus veitchii Hort.   PANDANACEAE
VEITCH SCREWPINE—Native to Polynesia. Tree, to 40', with stilt-like prop roots;
branches few, ending in huge clusters of strap-like leaves up to 8' long, and 3" or more
wide, bordered on both edges with white stripe, spiny-margined and with long-
pointed tip. Non-fruiting in Florida. Leaves used for mats, screens, and lamp shades.
Much grown as a pot plant in the North. Propagated by suckers or cuttings. Many
huge clumps on old properties have disappeared with land-clearing and new construction.

Parkinsonia aculeata Linn.   LEGUMINOSAE
JERUSALEM THORN; PALO VERDE—Considered probably native to tropical
America. Tree, to 30', with slender, green trunk and thorny, gracefully drooping
branches; foliage delicate, feathery; leaflets numerous, very small, produced on flat,
clustered leafstems to 16" long; flowers (spring-summer) yellow, 1" across, fragrant;
fruit a slim pod to 6" long. Propagated by seed, cuttings or air-layers. Drought- and
salt-tolerant; fast-growing in full sun.

Parmentiera cerifera Seem.   BIGNONIACEAE
CANDLE TREE; PANAMA CANDLE TREE—Native to Panama. Tree, to 20';
leaves evergreen, having 3 thin, pointed leaflets to 2" long; flowers to 2½" long,
bell-shaped, whitish with brown calyx; fruits candle-like, usually 1' but may be 4'
long and up to 1" thick, light-yellow, waxy, borne in abundance under favorable
conditions; relished by livestock. Grown from seeds. Rare in South Florida.

Passiflora coccinea Aubl.   PASSIFLORACEAE
RED PASSION FLOWER—Native to northern South America. Vine, rusty-hairy;
leaves oblong, wavy-edged, to 6" long and 2¾" wide, downy beneath; flower brilliant
scarlet with white-pink-and-purple corona, to 5" wide, lasting only one day; fruit
nearly round, 2" wide, with acid pulp and many minute seeds. Introduced from
Bolivia by Prof. Ira Nelson in 1954. Seeds distributed to members by Louisiana
Society for Horticultural Research in 1956. Now sold by South Florida nurseries.

Passiflora edulis Sims.   PASSIFLORACEAE
PURPLE PASSIONFRUIT—Native to southern Brazil. Vine, woody, perennial,
climbing by tendrils; leaves evergreen, glossy, deeply 3-lobed, finely toothed, to 8"
long; flowers fragrant, to 3" wide, white, with purple-and-white corona; open at noon
and close at nightfall; fruit nearly round, to 2½" wide, with dark-purple rind, yellow,
juicy pulp and many seeds. The YELLOW PASSIONFRUIT (*P. edulis* f. *flavicarpa*
Deg.) has larger flowers, yellow fruit; succeeds better in South Florida while the Purple is hardy and fruitful in Pasco County. Both yield delicious juice.

Pedilanthus tithymaloides Poit.   EUPHORBIACEAE
SLIPPERFLOWER; REDBIRD CACTUS; DEVIL'S BACKBONE—Native to
Central America, northern South America, also West Indies. Shrub, to 6', with succulent, green stems, frequently zigzag, containing milky sap; leaves dark-green,
triangular with tapering tip, up to 5" long; flowers tiny, enclosed in slipper-like, red
or purple bracts up to ¾" long. Grown from cuttings. Commonly cultivated varieties
have leaves variegated with white. Sap irritating to skin and eyes.

Jerusalem Thorn—*Parkinsonia aculeata*

Copperpod—*Peltophorum pterocarpum*

Garlic Vine—*Pseudocalymma alliaceum*

Ricasol Pandorea
Podranea ricasoliana

Traveler's Tree (dry fruit cluster)
Ravenala madagascariensis

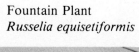

Fountain Plant
Russelia equisetiformis

Rangoon Creeper—*Quisqualis indica*

Mexican Flame Vine
Senecio confusus

Purple Passionfruit—*Passiflora edulis*

Slipperflower—*Pedilanthus tithymaloides*

Peltophorum pterocarpum Backer (*P. inerme* Llanos)   LEGUMINOSAE
COPPERPOD, or "YELLOW POINCIANA"—Native from India to Australia.
Tree, to 80', with umbrella top; leaves deciduous, to 1 ½' long, twice compound, finely divided; leaflets oblong, ½" long, pale on the underside; flowers (May-August)
fragrant, bright-yellow, 1 ½" wide, with 5 wavy and crinkled petals, borne in erect,
branched panicles to 1 ½' long, the flower stalk and buds densely coated with brown
hairs; seed pod maroon, flat, pointed, smooth, to 4" long, containing 1 to 5 hard
seeds which are slow to germinate unless pre-treated. Fast-growing, blooming in 4
years from seed; drought- and salt-tolerant; an excellent ornamental tree except for
surface roots; also is easily blown over in windstorms.

Pentas lanceolata Deflers   RUBIACEAE
EGYPTIAN STARCLUSTERS—Native to tropical Africa and Arabia. Herb, or
shrub, to 5', hairy-stemmed; leaves oval, pointed, up to 6" long, with indented veins
giving an attractive "corduroy" effect, slightly downy; flowers 5-pointed, star-like,
½" wide on ⅝" tube, numerous, in upright, round-topped clusters, lavender, pink,
rose-red or white. Grown from stem cuttings and blooms all year in full sun or partial
shade. Needs watering and cutting back occasionally.

Peperomia magnoliifolia A. Dietr. (*P. obtusifolia* sensu DC. & Urb. NOT A. Dietr.)
PEPEROMIACEAE (formerly PIPERACEAE)
OVALLEAF PEPEROMIA—Native from southern Mexico to northern South
America and naturalized in wet areas of South Florida. Herb, with succulent stems
to 2' long, trailing and rooting at nodes; leaves evergreen, alternate, variable,
usually broad-oval, dark-green, sometimes variegated with pale-yellow, glossy, 2 to
6" long; flowers minute in erect green spikes to 5" long. Easily grown from cuttings
in partial shade; popular as a ground cover and pot plant; important as a commercial "foliage" plant in Florida.

Persea americana Mill. (*P. gratissima* Gaertn.)   LAURACEAE
AVOCADO; ALLIGATOR PEAR—Native to tropical America. Tree, to 60', upright or spreading; leaves evergreen, to 10" long and 6" wide, usually long-oval,
variable, clustered; flowers small, greenish, in clusters; fruit up to 7" wide and 15"
long, dark-green to dark-purple, smooth or rough-skinned, nearly round, more often
pear-shaped, sometimes with elongated neck; flesh pale- to rich-yellow or greenish-
yellow, buttery, some with "nutty" flavor; cavity contains large, brown seed.
Propagated by budding and grafting. Grown as home fruit and commercially in
South Florida; many named varieties, especially of the West Indian race and hybrids
between it and the Guatemalan. The MEXICAN AVOCADO (*P. drymifolia* Schl. &
Cham.), cultivated in California, is rarely seen in Florida.

Avocado
Persea americana

Philodendron selloum

Petrea volubilis Jacq.   VERBENACEAE
QUEEN'S WREATH; PURPLE WREATH—Native to West Indies and northern
South America. Vine, woody, climbing; leaves pointed-oval, to 8" long, rough;
flowers (spring-summer or longer) purple, 5-petaled and enhanced by 5-pointed, star-
like, bluish-purple calyx up to 1 ½" wide; in elongated, drooping clusters 8" or more
in length; petals fall, leaving the calyx which is long-lasting but fades with age. Stems
sometimes used as ropes. Suckers are transplantable and branch tips can be ground-
layered. Fairly popular in South Florida.

Philodendron spp.   ARACEAE
Native to tropical America. Vines, or shrubs, valued for their lush green leaves, and
ranging from the small Heartleaf, *P. oxycardium* Schott (*P. cordatum* Hort.), grown
as a hanging houseplant, to the tree-climbing GIANTLEAF, *P. giganteum* Schott,
with stout, hairy stem and leaves up to 2' long and 1 ½' wide. Another, *P. erubescens*
Koch & Aug., is a moderately heavy tree-climber with triangular leaves to 1 ½' long,
reddish-purple beneath, and with a red sheath covering the buds of its white flowers.
P. radiatum Schott, known in the nursery trade as *P. dubium,* has leaves which stand
out stiffly on long stems and are also to 1 ½' in length but deeply cut into many
narrow segments. Grown from cuttings. Most thrive in indirect light, but *P. selloum*
C. Koch, the most common of the non-climbing, or so-called "self-heading", species
needs full sun. It has a thick, erect stem with supporting aerial roots and deeply cut
leaves to 3 ½' long. Hand-pollination yields seeds for propagation.

Canary Date Palm–*Phoenix canariensis*

Phoenix canariensis Hort.　　　　　　　　　　　　　　　　PALMAE
CANARY DATE PALM—Native to Canary Islands. Palm tree, to 60'; trunk criss-crossed with bases of old leaf-sheaths; does not sucker; leaves feather-shaped, up to 15' long, with greenish-yellow stems bearing numerous long, sharp spines at the base; fruit oval, ¾" long, golden-yellow, edible but of poor quality. Grown from seed. A superb, wind-resistant, drought- and salt-tolerant palm, valued for home and street landscaping.

Phoenix dactylifera Linn.　　　　　　　　　　　　　　　　PALMAE
DATE—Native to North Africa. Palm tree, to 100'; trunk criss-crossed with bases of old leaf-sheaths; suckers spring up around base of tree; leaves feather-shaped, to 20' long and 3' wide, spiny near base; fruit oblong, up to 3" long, hanging in several multiple-stemmed clusters. Sometimes of good quality in Florida, especially if picked underripe and carefully cured. Best grown from suckers.

Phoenix reclinata Jacq. (*P. senegalensis* Van Houtte)　　　　PALMAE
SENEGAL DATE—Native to Africa. Palm tree, to 25'; trunk rough, usually several rising from suckers, forming dense clumps; leaves feather-shaped, spiny at the base, stiffly curved, up to 7' long and 3' wide, white-hairy beneath, particularly when young; fruit oval, red or brown, ¾" long, edible but inferior. Propagated by seed and transplanting of suckers.

Pigmy Date Palm—*Phoenix roebelinii*

Phoenix roebelinii O'Brien   PALMAE
PIGMY DATE PALM—Native to Burma and Cochin China. Dwarf palm tree, to 10' or more; trunk rough, up to 6" in diameter (sometimes multiple), topped by a crown of feather-shaped, arching leaves to 4' long; leaflets slim, dark-green, with silky sheen; fruit deep-red, ½" long. Slow-growing, from seeds which remain viable 2-3 months and germinate in 5 weeks. An invaluable palm for Florida dooryards; also popular as a potted palm.

Phoenix rupicola T. Anders.   PALMAE
CLIFF DATE—Native to India. Palm tree, to 25'; trunk slender, smooth; leaves feather-shaped, up to 10' long, bright-green, not stiff, drooping; fruit cylindrical, yellow, ¾" long. A very desirable, graceful palm; rare in South Florida but worthy of cultivation in dooryards.

Phyllanthus acidus Skeels (*Cicca distichus* Linn.)   EUPHORBIACEAE
OTAHEITE GOOSEBERRY; STAR GOOSEBERRY—Native to India and Madagascar. Tree, to 20', with spreading main branches and clusters of slim foliage-branches bearing alternate leaves, oval, pointed, up to 3" long, soft, bright-green, the new growth rose-tinted; flowers small, yellow or reddish; fruit hangs in clusters along the branches, rounded, flattened, 6- or 8-lobed, pale-yellow, waxy-skinned, with crisp, translucent flesh, highly acid, makes ruby-red jelly. Cultivated in South Florida as a rare fruit tree and sometimes found as an escape.

[117]

Aluminum Plant—*Pilea cadierei*   South Florida Slash Pine—*Pinus elliottii*

Pilea cadierei Gagnep. & Guill.   URTICACEAE
ALUMINUM PLANT—Native to southeast Asia. Herb, succulent, to 1½' high; leaves opposite, ovate, toothed near the tip, to 3½" long, bluish-green with silver streaks. Fast-growing from cuttings in moist soil and partial shade. Requires attention until well established and replacement when straggly. Introduced from Belgium by Robert Wilson in 1951. *P. cadierei minima* is a pink-stemmed, dwarf cultivar.

Pilea microphylla Liebm.   URTICACEAE
ARTILLERY PLANT—Native from southern Mexico to northern South America and the West Indies. Herb, annual, succulent, ferny, erect, to 15" or reclining; leaves fleshy, more or less oblong-oval, to ⅜" long, light-green; flowers minute, greenish-white-and-red, male and female in same or separate clusters, the male ejecting pollen with force when jostled. A popular edging plant or ground cover for moist, shady locations, but apt to become a pest. Even small fragments quickly take root.

Pinus elliottii var. *densa* Little & Dorman   PINACEAE
SOUTH FLORIDA SLASH PINE—Native to southern Florida and some of the Keys and northward on the coasts of central Florida. Tree, to 100', normally with dense, symmetrical head; in South Florida usually unbranched and bare-trunked except near the top; needles to a foot long, dark-green, glossy; cone up to 7" long, carrot-shaped before opening to release seeds; with a short, sharp spine near tip of each scale. Wood exceedingly hard, used in many of the older South Florida homes. Fast-growing from seed. Cannot stand traffic above root system. Trees kept in landscaping schemes where mowing and other activity takes place will sooner or later die, one by one.

Piscidia piscipula Sarg. (*P. communis* Harms;   LEGUMINOSAE
    Ichthyomethia piscipula A. Hitchc.)
JAMAICA DOGWOOD; FISHFUDDLE TREE; FISH-POISON TREE—Native to South Florida, the Bahamas and West Indies. Tree, to 50'; bark gray, scaly; rounded head; leaves evergreen, pinnate with grayish-green, oval leaflets to 11" long; leafbuds and newly opened leaves satiny reddish-brown; flowers pea-like, white-and-lavender; ¾" long in elongated clusters; fruit a 4" light-brown pod with 4 papery,

Jamaica Dogwood—*Piscida piscipula*

scalloped wings; contains several red-brown, flat seeds. Bark, leaves, roots and twigs used to stupefy fish. A common tree of the Florida Keys.

Pithecellobium dulce Benth. LEGUMINOSAE
  (Genus formerly spelled *Pithecolobium*)
MANILA TAMARIND; GUAMUCHIL—Native to Mexico and Philippine Islands. Tree, to 50', with spreading head; spines in pairs on twigs; foliage evergreen, new growth reddish; leaflets in pairs, up to 2" long; flowers whitish, small, fluffy, in rounded clusters; fruit a loosely coiled, more or less red, pod ½" wide and 4 to 8" long, filled with black seeds, each with a thick, sweet, edible, pulpy aril, usually white, rarely red. Formerly much planted as a street tree in South Florida but found highly susceptible to hurricane damage and otherwise undesirable. Many still standing, and thorny saplings spring up from seeds dropped by birds.

Pittosporum tobira Dry. PITTOSPORACEAE
TOBIRA; JAPANESE PITTOSPORUM—Native to China and Japan. Shrub, to 10'; leaves evergreen, up to 4" long, leathery, dark-green; flowers 5-petaled, white, fragrant; fruit angular, ½" long and covered with short hairs. Grown from seed or cuttings. Withstands salt spray in seaside gardens. Variety *variegatum* Hort. has leaves marked with white.

Platycerium bifurcatum C. Chr. (*P. alcicorne* Desv.) POLYPODIACEAE
STAG-HORN FERN; ELK'S-HORN FERN. (Not to be confused with *Polypodium punctatum* var. *cristatum* Hort., q.v., which is known by the same common names.)—Native to Australia. A handsome fern, usually grown on a piece of moss-covered wood or section of tree-fern trunk suspended like a hanging basket or attached to a shaded wall; grows more vigorously when nailed directly to a tree. The center is 8 or 10" across, cabbage-like, with tightly overlapping fronds; extending out 2 to 3' or more on either side, are antler-like, forked fronds, downy on underside

Stag-Horn Fern—*Platycerium bifurcatum*

when young. Propagated by spores or division; needs semi-shade and adequate moisture.

Pleomele fragrans Salisb. (*Dracaena fragrans* Ker.-Gawl)   AGAVACEAE
FRAGRANT DRACAENA—Native to West Africa. Herbaceous plant, to 20';
leaves up to 3' long and 4" wide, curving out and downward, glossy, green, or green
striped with white or yellow; flowers open and distinctly fragrant after 5 P.M., small,
yellow, in many globular clusters on a branched inflorescence 1' or more long; fruit a
scarlet, ill-smelling berry. Grown from seed or pieces of stem cut 4" to 6' in length.
Does best in shade.

Fragrant Dracaena—*Pleomele fragrans*   *Pleomele reflexa*

Pleomele marginata N.E.Br. (*Dracaena marginata* Lam.)   AGAVACEAE
Native to Madagascar. Shrub, to 12'; stem slender, scarred, topped by cluster of narrow, sword-shaped leaves up to 2' long, with sharp point at tip, green with purple edges; flowers in long spikes. Slow-growing from cuttings or air-layers. Popular as a patio or foundation shrub and indoor plant.

Pleomele reflexa N.E. Br. (*Dracaena reflexa* Lam.)   AGAVACEAE
Native to India, Malaysia, Madagascar and Mauritius. Shrub, with flexible, branching stem to 12' high, densely set with clasping, evergreen, drooping leaves—oblong, pointed, glossy, leathery, to 5½" long. There are narrow-leaved, also variegated, forms. Very easily grown from cuttings. In recent years, popularized as a house and patio plant. Inclined to droop and often requires tying back or staking but otherwise highly desirable.

Plumbago auriculata Lam. (*P. capensis* Thunb.)   PLUMBAGINACEAE
CAPE PLUMBAGO; LEADWORT—Native to South Africa. Shrub, herb-like, with very slim stems, forming a dense mass up to 4' high or may climb on a support; leaves evergreen, to 3" long, slender, pointed, often in clusters of 3; flowers 5-petaled, pale-blue, about ½" across at the end of a slim tube protruding from a sticky-haired calyx. Variety *alba* Hort. has white flowers. The plant contains a poisonous principle and has many medicinal uses. Grown from seed and cuttings. Popular in foundation plantings of South Florida homes.

Cape Plumbago—*Plumbago auriculata*

Plumeria alba Linn. (Genus sometimes spelled *Plumiera*)   APOCYNACEAE
WHITE FRANGIPANI—Native to West Indies. Tree, to 35', with spreading top; leaves deciduous, to 12" long and 3" wide, long-pointed, downy on underside; flowers 1" wide, white with yellow center, fragrant, clustered. Fast-growing from cuttings.

Plumeria emarginata Griseb.   APOCYNACEAE
BLUNT-LEAVED FRANGIPANI—Native to West Indies. Tree, to 20', with slender trunk and few spreading branches; leaves deciduous, oblong, with horizontal veins, leathery, notched at tip, to 8" long and 3" wide, clustered at ends of stout twigs; flowers white with yellow center, up to 2" across, in clusters, fragrant. Grows fast from cuttings.

Plumeria rubra Linn.   APOCYNACEAE
NOSEGAY FRANGIPANI; ROSE-PINK JASMINE—Native from Mexico to northern South America and West Indies. Tree, to 15', with white, gummy sap, slim trunk and only a few, spreading, main branches triple-forked like chicken feet at the tip, forming a parasol-like top; leaves deciduous, handsome, to 18" long and 6" wide, rounded or short-pointed at the apex, in rosette-clusters at the ends of the stubby branchlets; flowers 2" wide, with rose-pink, coral-pink or, rarely, wine-red, petals, usually yellow-throated, in lovely, prominent clusters. Tree totally bare in late winter yet picturesque in attitude. Large cuttings may be dried and then set directly in the ground.

Plumeria rubra var. *acutifolia* Bailey   APOCYNACEAE
  (*P. acuminata* Ait.; *P. acutifolia* Poir.)
MEXICAN FRANGIPANI; TEMPLE FLOWER; GRAVEYARD FLOWER—Native to Mexico. Tree, to 30', with broad head; leaves deciduous, lance-shaped, 3" wide and up to 15" long, in clusters at ends of branches, shed in late winter; flowers very fragrant, about 2" across, in spreading clusters, with white or pale-yellow petals and yellow centers; seed pod (rare) to 7" long and ¾" wide, borne in pairs end-to-end, containing winged seeds. Milky latex of this and other species purgative in quantity and irritating if left on the skin; used, like other parts of the tree, in folk remedies. Flowers yield frangipani perfume. Grown from seeds or large cuttings. Dried before planting.

Nosegay Frangipani–*Plumeria rubra*

Podocarpus gracilior Pilg.   PODOCARPACEAE
Native to East Africa. Tree, to 100' in its place of origin; leaves evergreen, alternate, at the ends of the drooping branches, very slim, pointed, glossy; male cones slender, to 1" long, in 2's or 3's; seed green or purplish with a bloom, to ¾" long. Recently being planted as a foundation shrub, bushy and of soft, graceful appearance. Propagated by cuttings and air-layers; does well in sun or shade.

Podocarpus macrophyllus var. *maki* Sieb.   PODOCARPACEAE
SHRUBBY YEW—Native to China and Japan. Tree, to 50' with upright branches; leaves evergreen, very slim, to 3" long; seed oval, to ½" long, green with bluish bloom, and attached to the outer end of a small, fleshy, edible, reddish-purple receptacle, the whole resembling in miniature the nut and "apple" of the Cashew. Grows fast from seed, cuttings or air-layers; usually trimmed to compact, columnar form as a foundation plant; also grown as a hedge.

Podranea ricasoliana Sprague   BIGNONIACEAE
  (*Pandorea ricasoliana* Baill.)
RICASOL PANDOREA—Native to South Africa. Climbing shrub; leaves deciduous, pinnate with 7 to 9 thin, pointed leaflets, tooth-edged and up to 2" long; flowers (spring) bell-shaped, 2" long, pale-pink with red stripes, hanging in loose clusters of as many as 20 blooms; fruit a slender pod up to 1' long. Branches easily ground-layered. Fairly common porch vine in South Florida. Needs full sun.

Polypodium polypodioides Watt   POLYPODIACEAE
RESURRECTION FERN—Native to South Florida, the Bahamas, West Indies and tropical America. Fern, with scaly, creeping rootstock; leaves erect, to 8" long, deeply divided into oblong segments, with numerous scales on the underside, dehydrating and curling up in dry weather, reviving and expanding when the atmosphere is moist. Abundant on limbs of oak trees and on rocks in hammocks. Easily grown on pieces of old wood in shade. Dried fern sold for medicinal use in Cuba and Puerto Rico.

Polypodium punctatum Sev. (syn. *P. pterocarpon* Cav.)   POLYPODIACEAE
  var. *cristatum* Hort.
STAG-HORN or ELK'S HORN FERN (Not to be confused with the quite different STAG-HORN or ELK'S HORN FERN, *Platycerium bifurcatum,* q.v.)—The species, of which this is a form, is native to the Old World tropics. Fern, up to 2½' tall, with vertical, light-green, flat fronds, narrow at base, flaring to 4" wide at top; top "crested", i.e., toothed and ruffled. Popular in plant boxes and in garden borders. Propagated by division.

Polyscias balfouriana Bailey (*Aralia balfouriana* Hort.)   ARALIACEAE
BALFOUR POLYSCIAS—Native to New Caledonia. Shrub, to 25', bushy, compact, with upright branches; leaves evergreen, compound, with 3 nearly circular or somewhat irregular leaflets up to 4" wide, tooth-edged, sometimes outlined with white and white-blotched; flowers tiny, in groups clustered to form a large spray. Grown from cuttings. Common hedge plant in Key West and occasional on the mainland.

Polyscias scutellaria   "Aralia"–*Polyscias guilfoylei*

Polyscias filicifolia Bailey (*Aralia filicifolia* Moore)   ARALIACEAE
"ANGELICA"; FERNLEAF POLYSCIAS—Native to Oceania. Shrub, to 8', with
purplish twigs; leaves evergreen, fern-like, finely cut and somewhat curly, forming a
compact mass; flowers tiny, in clustered groups. Grown from cuttings. Common
hedge plant in Key West and gaining popularity as a pot and patio plant throughout
South Florida.

Polyscias guilfoylei Bailey (*Aralia guilfoylei* Bull.)   ARALIACEAE
GUILFOYLE POLYSCIAS—Native to South Sea Islands. Shrub, to 20', erect with
upward-growing branches; compact; leaves evergreen, pinnate, leaflets up to 6" long,
glossy, bristle-toothed, green or green bordered with white or with irregular white
patches. Seldom blooms. Grown from cuttings as a back-fence hedge in South
Florida. Excessive contact may produce dermatitis.

Polyscias scutellaria Fosb. (*Nothopanax scutellarium* Merr.)   ARALIACEAE
Place of origin unknown. Shrub, or small tree, slender, erect, to 12'; leaves simple or
composed of 2 or 3 leaflets, saucer-like, heart-shaped at base, nearly circular with
curled-up edge, very glossy on both sides, to 10" wide. Easily grown from cuttings. A
handsome, choice plant for patios still not sufficiently known. Grown in hedges in
Java.

Pongamia pinnata Pierre (*Derris indica* Bennet)   LEGUMINOSAE
PONGAM; POONGA OIL TREE—Native to tropical Asia, Africa, Australia,
Polynesia. Tree, to 75', with broad "umbrella" head and somewhat drooping
branches; leaves briefly deciduous, pinnate with leaflets to 4" long, bright-green,
glossy; flowers (spring) pink or lavender to white, pea-like in slim, pendent clusters;
fruit a flat, brown pod about 1 ½" long, containing a single, flat, circular seed. Seeds
medicinal, used to stupefy fish, as are the roots; seeds yield oil for illumination and
medicinal purposes; leaves, flowers and bark also medicinal. Fast-growing from seed;
wind-, salt- and drought-tolerant. Has been recommended as a shade tree for South
Florida but litters the ground with old leaves and pods.

[124]

Pongam–*Pongamia pinnata*

Christmas Vine–*Porana paniculata*

Porana paniculata Roxb.   CONVOLVULACEAE
CHRISTMAS VINE, BRIDAL VEIL—Native to India. Vine, climbing to 30',
slender-stemmed; leaves evergreen, attractively heart-shaped, pointed, up to 6" long,
with conspicuous veins, downy-white on underside; flowers (October-December)
funnel-like, white, 1/3" long and wide, profusely borne in elongated sprays, fragrant.
Seed capsule round, 3/16" wide with propeller-like sepals but rarely produced.
Propagated by stem cuttings. Vine will shroud trees if allowed to spread; favored for
Christmas decoration.

Pouteria campechiana Baehni (*Lucuma nervosa* DC.)   SAPOTACEAE
CANISTEL; EGGFRUIT; TI-ES—Native to northern South America. Tree, to 25';
leaves evergreen, slender, lance-shaped, pointed and gracefully curving, to 8" long,
clustered, bright-green, glossy; flowers whitish, inconspicuous; fruit (winter) almost
round with a beaked tip, or broadly oval and pointed, or oblong, up to 5" long and
3½" wide; smooth yellow skin and yellow, pasty pulp, musky and sweet, often
likened to yolk of hard-boiled egg. Seeds, 2 or 3, to 2½" long, glossy, brown with a
white side. Fast-growing from seeds; selected types grafted. Usually included in
tropical fruit tree collections in South Florida.

Pouteria sapota H. E. Moore & Stearn (*P. mammosa* Cronq.;   SAPOTACEAE
Calocarpum sapota Merr.; *Lucuma mammosa* Gaertn.)
SAPOTE; MARMALADE FRUIT; MAMEY COLORADO (Not to be confused
with the MAMEY or MAMEY DE SANTO DOMINGO, *Mammea
americana*)—Native to Central America. Tree, to 80' or more, upright, with handsome, symmetrical head; branches open and tipped with striking clusters of oval,
pointed leaves, to 12" long and 4" wide, whitish or brownish on the underside;
flowers small, white clustered; fruit oval, pointed, up to 8" long with rough, brown
skin; flesh salmon or orange-red, soft, sweet, somewhat pumpkin-like in flavor.
Formerly rare in South Florida, now in demand by new Cuban residents. Grafting
techniques being developed.

[125]

Pritchardia pacifica Seem. & H. Wendl.   PALMAE
PACIFIC PALM; FIJI FAN PALM—Native to Samoa and Fiji Islands. Palm tree, to 30'; leaves fan-like, stiff, light-green, about 3' wide and 4' long, on 3' stems, flaring out from top of trunk; flowers yellow, in tassel-like cluster hanging at end of thick, 3' stalk, the whole resembling a camel's tail; fruit black, ½" across. Seeds remain viable 4-6 weeks; germinate readily. Some specimens have been victims of lethal yellowing.

Pritchardia thurstonii F. Muell. & Drude   PALMAE
THURSTON PRITCHARDIA PALM—Native to Fiji Islands. Palm tree, to 15'; leaves fan-like, stiff, waxy and somewhat scaly on underside; flower stalk as much as 10' in length; fruits ¼" across in compact clusters. Grown from seed; somewhat hardier than *P. pacifica.*

Prunus persica Sieb. & Zucc.   ROSACEAE
PEACH—Native to South China. Tree, to 10', slender, willowy, with slim, curving, deciduous leaves; flowers up to 1" across, pink. Red Ceylon, a tropical variety, has oval, beaked fruit with fuzzy, greenish-yellow skin, red-blushed, and whitish flesh, strawberry-red in center, freestone; excellent raw, cooked or preserved. Commonly grown from seed for home use in South Florida; however, during the past few years, the fruit has been largely ruined by the Caribbean fruit fly.

Pseuderanthemum atropurpureum Radlk.   ACANTHACEAE
CAFE CON LECHE—Considered native to Polynesia. Shrub, to 6'; leaves broad-oval, up to 6" long, dark reddish-purple or lighter, or sometimes green or green-and-yellow; flower 1" wide, 5-lobed, white with purple eye and flecks, or all purple; fruit a club-shaped seed pod. Grown from cuttings.

Pseudobombax ellipticum Dugand   BOMBACACEAE
   (formerly *Pachira fastuosa* Decne.)
SHAVINGBRUSH TREE—Native to tropical America and the West Indies. Tree, to 30', with smooth, green bark and open, spreading branches; leaves deciduous, palmate, with 5 broad-oval leaflets to 9" long, red and downy when young; flowers, appearing in spring when tree is bare, solitary. The 5 petals, 3 to 5" long, purple outside, white and downy inside, curl back away from the large, showy tuft of long stamens, usually pink, occasionally white. Seed pod (produced only if flowers hand-pollinated in Florida) 4 to 6" long, splits into 5 sections containing tan wool and many, ¼"-long, brown seeds. Fast-growing from air-layers or large cuttings which root readily.

Pseudocalymma alliaceum Sandw. (long erroneously known   BIGNONIACEAE
   as *Cydista aequinoctialis* Miers)
GARLIC VINE—Native to West Indies and northern South America. Vine, climbing; leaves divided into 2 leaflets, often with a tendril between; leaflets to 4" long, glossy, have strong garlic odor; flowers in large, lovely clusters, bell-shaped, up to 3" long, pink, lavender, or white with rose or purplish veins; fruit a slender pod to 15" long. Grown from cuttings in full sun. Blooms off and on in warm months.

Pseudophoenix sargentii H. Wendl.   PALMAE
BUCCANEER PALM; HOG CABBAGE PALM; CHERRY PALM—Native to
two Florida Keys and Bahamas, Dominican Republic and Yucatan. Palm tree, 10 to
15' as usually seen but may reach 25'; trunk light-gray, smooth, ringed; leaves feather-
shaped, gray-green, to 7' long and about 3' wide; flowers yellowish, in branched
cluster to 3' long; fruit orange-red, round or 2- or 3-lobed, up to ¾" across, used by
early settlers for fattening hogs. Terminal bud edible. Slow-growing from seed. The
WINE PALM (*P. vinifera* Becc.) from Hispaniola is similar but taller. Sap formerly
made into alcoholic beverage. Seeds germinate in about one month.

Psidium cattleianum Sabine (*P. littorale* Raddi)   MYRTACEAE
CATTLEY GUAVA; STRAWBERRY GUAVA—Native to Brazil. Shrub or
slender tree, to 20', bark smooth, light-brown; leaves evergreen, to 4" long, dark-
green, smooth, leathery, glossy; flowers up to ¾" across, white with prominent
stamens; fruit (summer) round, to 2" across, with dark-red skin and white flesh composed of a thin outer layer surrounding soft central pulp which is filled with small
seeds. Flavor spicy, subacid, somewhat strawberry-like. Eaten fresh or preserved.
Variety *lucidum* Hort., YELLOW CATTLEY GUAVA, has clear yellow fruits
which are possibly even spicier. Grown from seed or air-layers.

Psidium guajava Linn.   MYRTACEAE
GUAVA—Native from Mexico to northern South America. Tree to 30', of slender
proportions but spreading; bark light-brown, scaly; leaves to 6" long, dull-green with
conspicuous veins; flowers white, up to 1½" across with brush of white-and-yellow
stamens; fruit round, oval or pear-shaped, to 6" long, usually highly odorous; skin
pale-yellow; flesh white, cream-colored, pink or reddish, somewhat granular, sur-
rounds softer central pulp which contains numerous small, bony seeds; flavor sour to
sweet, musky. Sour types best for jelly; improved sweet types excellent for dessert,
cooking and preserving. Leaves, bark and roots medicinal. Commonly occurs as an
escape from cultivation. Grown in South Florida for home use and to some extent
commercially. Bears more or less continuously but main crop is in August and
September. Superior types easily propagated by air-layering. In past few years, most
fruits are infested with larvae of Caribbean fruit fly.

Psychotria undata Jacq. (*P. nervosa* Sw.)   RUBIACEAE
WILD COFFEE; SEMINOLE BALSAMO—Native to South Florida, the Bahamas
and West Indies. Shrub or tree, to 15'; leaves evergreen, to 6" long, bright-green,
shiny, with prominent veins; flowers white, 3/16" wide, clustered; fruit scarlet, oval,
up to ⅜" long, with 2 seeds like coffee beans but not similarly used. Fruits apparently
harmless but characterless and relished only by birds. Easily grown from seed in sun
or shade.

Wild Coffee
Psychotria undata

Solitaire Palm—*Ptychosperma elegans*

Ptychosperma elegans Blume (*Seaforthia elegans* R. Br.,   **PALMAE**
   erroneously *Hydriastele wendlandiana*)
SOLITAIRE PALM; often erroneously called ALEXANDER'S PALM—Native to
Queensland, Australia. Palm tree, to 20', with straight trunk to 4" in diameter,
swollen at base, ringed, topped by slender crownshaft and 6-9 feather-like leaves, 3'
or more long on foot-long stems; leaflets 2½' long and 2-3½" wide; flowers white, in
bushy cluster to 1½' long borne well below the crownshaft; fruit red, rounded or
oblong, up to ¾" long. Seed remains viable 4-6 weeks. Much used in commercial
landscaping where space is limited, as well as in dooryards.

Ptychosperma Macarthuri H. Wendl.   **PALMAE**
   (*Actinophloeus Macarthuri* Becc.)
MACARTHUR PALM—Native to New Guinea. Palm with multiple, slender, ring-
ed stems in clumps, to 30'; leaves feather-like, to 7' long, with leaflets 1 to 2" wide and
to 1' long, oblique and jagged at the tip; flowers faintly fragrant, white or yellowish,
male and female alternating in branched clusters to 15" long; fruit round, orange-red,
3/16" long, in showy sprays. Fast-growing from seed in partial shade.

Punica granatum Linn.   PUNICACEAE
POMEGRANATE—Native from Persia to southern Asia. Shrub or tree, to 20',
often spiny; leaves deciduous, slender, to 4" long; flowers to 1 ½" wide, bell-like, with
bright orange-scarlet petals protruding from waxy, tubular, orange calyx and surrounding numerous yellow-tipped stamens; fruit round with floral remnant
protruding at apex, up to 5" in diameter; rind yellow with pink blush or deep-red,
leathery. Fruit filled with masses of pinkish to red, pulpy juice-sacs, each containing a
small seed. Juice-sacs refreshing; juice also expressed for use fresh or made into wine.
Grown from dormant cuttings as a dooryard ornamental; fruit does not equal the
quality of pomegranates produced in dry climates.

Pyracantha coccinea Roem.   ROSACEAE
RED FIRETHORN—Native to southern Europe and western Asia. Shrub, to 12' or
more, with spiny branches; leaves, narrow, to 2" long, finely toothed, often downy
below when young; flowers white, 1/3" wide, in compact clusters; fruit red, flattened-
globular, ¼" wide, in spectacular masses. Fast-growing from cuttings, difficult to
transplant; may be cultivated in South Florida but much more common from Central
Florida northward. "Tiny Tim" is a dwarf, nearly thornless hybrid introduced from
California in 1964.

Pyracantha koidzumii Rehd. (*Cotoneaster formosana* Hayata)   ROSACEAE
FORMOSA FIRETHORN—Native to Formosa. Shrub, to 12' or more, thorny;
leaves evergreen, oval, notched at tip, up to 2" long, clustered, light-green with a
bloom on underside; flowers white, delicate, ¼" across; fruit rounded, flattened, ⅜"
wide, orange-red to red, in compact clusters, abundant and showy in fall and winter.
This is the most common species in the southern U.S. Cultivars of *P. crenulata* Roem.
and *P. angustifolia* Schneid. are also being grown in South Florida. All grow rapidly
from seed or cuttings in sun and cover much space unless cut back in spring; may be
kept compact by espaliering.

Pyrostegia ignea Presl. (*Bignonia venusta* Ker.)   BIGNONIACEAE
FLAME VINE—Native to Brazil. Vine, climbing; leaflets in 2's or 3's, one sometimes taking form of a 3-parted tendril, the others being pointed-oval and up to 3"
long; flowers (January-April) bright-orange, to 3" long, tubular with 5 points curling back from the mouth, borne many in a cluster, showy. In Brazil it bears pods 1'
long. Fast-growing from stem cuttings. One of the most spectacular vines of Florida
gardens, blanketing roofs and walls or draping tall trees. Should be pruned severely
after blooming.

Quamoclit pennata Bojer (*Q. quamoclit* Britt.)   CONVOLVULACEAE
CYPRESS-VINE; INDIAN PINK—Native to tropical America and West Indies;
naturalized in Florida and west to Texas. Vine, annual, to 20', slender-stemmed;
leaves finely feathery, up to 7" long; flowers scarlet, 1 ½" long, funnel-shaped, the
flare 5-pointed, single or in clusters; seed capsule oval, ½" long, containing smooth,
black seeds. Hard seeds must be notched or soaked to hasten germination. Often
planted as an ornamental. Prefers sunny location.

Live Oak—*Quercus virginiana*

Quercus virginiana Mill.   FAGACEAE
LIVE OAK—Native from Virginia to southern Florida and Cuba and along the Gulf Coast to Mexico. Tree, to 50' with thick trunk and wide-spreading branches; leaves nearly evergreen but shed copiously in winter in South Florida, alternate, leathery, 2 to 5" long and ½" to 2½" wide, dark-green above, pale beneath; flowers (spring) tiny, male in drooping catkins to 3" long, female, solitary or grouped at leaf bases; acorn, oval, smooth, ¾" long, with rough cup; kernel oily, edible. This magnificent native oak is not slow-growing as commonly thought but fairly fast from seed if watered and fertilized. It is also easily transplanted and should be more commonly utilized as a shade and street tree.

Quisqualis indica Linn.   COMBRETACEAE
RANGOON CREEPER—Native to southern Asia and East Indies. Climbing shrub, to 35', woody, with twining stems becoming thorny with age; leaves evergreen, oblong-oval, pointed, to 6" long, brownish fuzz on new growth; flowers (summer) 1" across, 5-petaled, with 2 to 3" calyx tube, in drooping clusters, white at first, then pink and finally red, or light-red when first open, turning to dark-red; fruit 1" long, slender, with 5 lengthwise wings; used medicinally when half-ripe. Ripe seeds taste like coconut and are used for flavoring but are poisonous in quantity; leaves medicinal; stems used for basketry. Fast-growing from cuttings or suckers; needs much space and strong support.

Randia aculeata Linn. (*R. mitis* Linn.)   RUBIACEAE
BOX BRIAR—Native to South Florida, the Bahamas, West Indies and Mexico. Shrub in Florida; elsewhere may be a tree to 20' with short, stiff, thorny branches; leaves to 2" long, shiny, in clusters; flowers white, up to ½" across, clustered, fragrant; fruit round or oval, ⅜" long, white. Used as a Christmas tree in the Virgin Islands; wood used for fashioning culinary implements in the Netherlands Antilles.

[130]

Rapanea guianensis Aubl.   MYRSINACEAE
GUIANA RAPANEA; MYRSINE—Native to South Florida, the Bahamas, West
Indies and northern South America. Shrub or tree, to 20'; leaves evergreen, lance-
shaped, to 5" long, clustered; flowers greenish, tiny, close along branches; fruit
round, up to 3/16" across, black, glossy, conspicuous but eaten only by birds.

Ravenala madagascariensis J.F. Gmel.   STRELITZIACEAE
TRAVELER'S TREE—Native to Madagascar. Herbaceous plant when young, with
banana-like leaves, up to 9' long, which rise from the base and spread out, giving the
entire plant the appearance of a fan; when old, the fan-like arrangement of leaves is
supported by a sturdy, palm-like trunk up to 40' high. Clear sap may be obtained
from bases of leafstalks to quench thirst and may be boiled down to sugar. Flower
cluster is an erect series of canoe-like bracts containing small white blossoms. Seeds
black, covered by indigo-blue arils which are shed on the ground. Seeds and oily arils
edible. A valued ornamental in South Florida, grown from seed or suckers.

Reynosia septentrionalis Urban   RHAMNACEAE
DARLING PLUM; RED IRONWOOD—Native to South Florida, the Bahamas
and West Indies. Shrub or tree, to 30'; bark red-brown, scaly; leaves evergreen, to
1 ½" long, leathery, notched at tip; flowers yellowish-green, small, clustered; fruit
oval or round, up to ¾" long, dark-purple, sweet, edible.

Rhaphidophora aurea Birdsey   ARACEAE
(*Epipremnum aureum* Bunt.; *Scindapsus aureus* Engl; *Pothos aureus* Lind.)
HUNTER'S ROBE; TARO VINE—Native to Solomon Islands. Climbing plant,
with strong, rope-like stem which puts out aerial roots; leaves evergreen, heart-
shaped, tapering to a point, up to 18″ long and 1 ′ wide, green, often with yellowish-
white patches. Familiar as a pot plant and commonly seen growing up the trunks of
royal palms and coconut palms. May be used as a ground cover in shady locations.
Grown from cuttings. Juice may irritate skin.

Rhapis excelsa Henry (*R. flabelliformis* Ait.)   PALMAE
LADY PALM—Native to China and Japan. Palm with creeping rootstock sending
up slender stems 1" wide and to 10' high, forming clumps; leaves 1' wide, fan-shaped
but deeply divided into 5 to 7, ribbed, strap-like segments 1 to 2" wide. Propagated by
offsets. Slow-growing, thrives best in shade. Admirable dwarf palm for hedges, foundation planting, patios and indoors. Peeled stems formerly made into walkingsticks.
R. humilis Blume is similar but its leaves have 7 to 10 segments. It is often said to be a
smaller palm than *R. excelsa* but in California is taller—to 18'.

Rhizophora mangle Linn.   RHIZOPHORACEAE
RED MANGROVE—Native to Florida, the Bahamas, West Indies and tropical
America. Tree, usually shrubby, may reach 70', with exposed stilt-roots arching out
from the base; tree forms dense thickets along coasts, and islands in shallow water;
leaves evergreen, leathery, to 6" long; flowers ¾" across, yellow, hairy; fruit conical,

Red Mangrove—*Rhizophora mangle*

1" long, brown, produces a green sprout to 1' in length that falls and takes root. These somewhat pencil-shaped sprouts, found on beaches, are used by children for drawing in the sand, are accordingly called "sand-pencils". This tree is not only picturesque but invaluable as a land-builder and retainer. Oysters often found clinging to roots. Bark widely used in tanning; bark, roots and leaves employed in folk medicine.

Rhoeo spathacea Stearn (*R. discolor* Hance; *Tradescantia*   COMMELINACEAE
    discolor L'Her.)
OYSTER PLANT; MOSES-IN-A-BOAT—Native to Mexico, Central America, West Indies. Herb, succulent, with very short stem and rosette of upright, fairly stiff, dagger-like leaves about 1½" wide and up to 1' long, green on top and purple on underside; flowers tiny, white, in a purple, boat-like sheath. A common, rapidly multiplying ground cover; will grow even in crevices of rock walls and on roofs. Easy to transplant; requires no care. Leaves eaten by ducks and raccoons but watery juice may irritate human skin.

Ricinus communis Linn.   EUPHORBIACEAE
CASTOR BEAN; CASTOR OIL PLANT—Considered probably native to Africa. Herbaceous plant, somewhat woody with age; up to 40', much branched; leaves evergreen, to 3' across, medallion-like, deeply cut, usually with 8 or 9 points, green or purplish-red; flowers yellow in an upright spike, followed by bur-like fruits, green, bluish or red, up to 1" across, which dry, turn brown and split open scattering the seeds. Seeds vary from ¼ to ½" in length and in color from white or tan mottled with dark-brown to all black; poisonous; yield castor oil. Planted in South Florida as a

Sarita—*Saritaea magnifica*

African Tulip Tree
Spathodea campanulata

Miracle Fruit—*Synsepalum dulcificum*

Pink Trumpet—*Tabebuia* hybrid

Yellow Elder—*Tecoma stans*

Bush Clockvine—*Thunbergia erecta*

Glorybush—*Tibouchina urvilleana*   Limeberry—*Triphasia trifolia*

possible source of lubricating oil during World War I; has become a pestiferous weed on undeveloped property. Unworthy of cultivation as an ornamental plant.

Rosa spp. ROSE   ROSACEAE
In older gardens of South Florida, one finds the so-called Florida Rose, a double, red climber of doubtful origin which blooms for many years. And in Key West and south Dade County, one also finds perennial types even in the smallest dooryards. These include the Key West Pink, Rock Rose (cream to pink), Red Rock (dark red), White Guava, Pink Guava, Blushing Bride (yellow flushed with pink), Miniature or Tom Thumb Rose, and the Etoile de Lyon (yellow). Some rose fanciers have collections of more than 100 old-time varieties requiring little care. Experimenting with the many new varieties offered by the trade is a popular hobby with South Florida rose lovers who formerly had to resort to dormant grafted plants chiefly from Texas which bloomed well for a year or two and were then replaced. The adoption of *Rosa fortuniana* as a dependable rootstock has made Florida-grafted rose bushes available and long-lived; and commercial rose production has flourished in the State since 1960.

Roystonea elata F. Harper (*R. floridana* O.F. Cook)   PALMAE
FLORIDA ROYAL PALM (Not *R. regia* O.F. Cook, the Cuban Royal Palm, formerly considered the same)—Native to Florida. Palm tree, to 125', with smooth, cylindrical, light-gray trunk topped by a sleek, green crownshaft up to 8' or more in height, and dark-green, feather-shaped leaves to 15' long and 6' wide, the leaflets having few if any lengthwise nerves; flowers ¼" wide, white, in hanging cluster up to 2' long; fruit round, ½" long, dark-purple, with one seed. Fruit and terminal bud edible; leaves used for thatching. *R. regia* reaches 70', the leaflets have conspicuous lengthwise nerves, the inflorescence is shorter and wider, and the fruit oval. Both palms are equally represented in Florida landscaping. Seeds viable only 4-6 weeks.

Castor Bean—*Ricinus communis*

Ruellia brittoniana Leonard　　　　　　　　　　　　　　　ACANTHACEAE
Native to Mexico; found wild in Florida. Herb, shrubby, to 3', leaves very slender, pointed and up to 12" long, sometimes a little wavy; flowers lavender-blue, tubular, flaring at apex, up to 1 ½ " long, in clusters. This plant has long been confused with *R. malacosperma* Greenman which apparently is not in cultivation. Useful bedding plant for damp, shady locations.

Russelia equisetiformis Schl. & Cham. (*R. juncea* Zucc.) SCROPHULARIACEAE
FOUNTAIN PLANT; FIRECRACKER PLANT; CORAL PLANT—Native to Mexico and Central America; naturalized in Florida and West Indies. Shrub, to 8', drooping, has an abundance of slim, wispy, green branches, usually leafless, or with a few tongue-shaped, ½"-long leaves on the larger stems; flowers (all year) bright-red, slender, tubular, about 1" long, in sprays. Grown from cuttings or ground layers; forms billowy masses in partial shade with little water.

Sabal palmetto Lodd.　　　　　　　　　　　　　　　　　　PALMAE
CABBAGE PALMETTO—Native to southeastern U. S. and is the state tree of Florida. Palm tree, to 80'; trunk criss-crossed with "boots" of old leaf-bases until old, when trunk is nearly smooth; leaves fan-shaped but slightly folded instead of out-spread, the recurved midrib extending about 1/3 of the distance from the base to the top and the narrow leaf-segments drooping; leaves up to 8' long, on stalks up to 7' long, form a round head; flowers ¼" wide in hanging cluster up to 6' long; fruit round, ½" wide, black, much eaten by the early Indians of Florida. The terminal bud has been commonly eaten as "palm cabbage" but the tree is now protected by State law. Trunks were used by Seminoles in the construction of shelters; have also served as piling for docks; fiber products are made from the leaves and leafstalks; unopened leaves are gathered and stripped for use in church services on Palm Sunday. This palm is increasingly used in coastal and inland landscaping; many have been successfully transplanted from the wild.

Samanea saman Merr. (*Pithecellobium saman* Benth.)　　　LEGUMINOSAE
RAIN TREE; SAMAN; MONKEY-POD—Native to Central America and West In-dies. Tree, to 80', with wide-spreading horizontal branches forming shallow, arched head; foliage deciduous, feathery; leaflets to 2" long, which close when rain threatens, are shed in late winter; flowers yellowish, brush-like, in fan-shaped clusters; fruit a narrow pod to 8" long, containing seeds embedded in sweet, brown pulp relished by livestock. Fast-growing from seed; rare in South Florida.

Sambucus simpsonii Rehd.　　　　　　　　　　　　　　SAMBUCACEAE
　　　　　　　　　　　　　　　　　　　　　　(formerly CAPRIFOLIACEAE)
FLORIDA ELDER—Native to Florida, West Indies and tropical America. Shrub or tree, to 15', with rough bark; leaves evergreen, pinnate with leaflets up to 3" long, glossy, toothed, pointed; flowers tiny, numerous, in broad, flat clusters, white, fragrant; fruit (summer) round, ⅛ to ¼" wide, nearly black, juicy, glossy, edible; good for pie, jam, jelly and wine. Grows abundantly in lowlands.

Ruellia brittoniana   Sanchezia—*Sanchezia nobilis*

Sanchezia nobilis Hook. f.   ACANTHACEAE
SANCHEZIA—Native to Ecuador. Shrub, scarcely woody, to 6'; leaves evergreen, to 18" long, oval, pointed at both ends; flowers tubular with rolled edge, 2" long, in upright spike, yellow, with red bracts. Variety usually grown has leaves variegated with yellow. Raised from cuttings or air-layers in partial shade or full sun, with ample watering.

Sansevieria spp.   AGAVACEAE
BOWSTRING HEMP—Native to Africa and Asia. Herbs, perennial, with stiff, vertical, ribbon-like leaves, up to 5' tall, grown as ornamental houseplants and in gardens because of their variegated coloring; usually have cross-bands of alternating dark- and light-green; some have a yellow, white or brown stripe on edges of leaves; flowers tubular, yellow or whitish in upright spike. S. *trifasciata* Prain is the common SNAKEPLANT, a useful tall ground cover for shady, rocky locations but self-multiplying and runs wild. Sansevierias yield fiber for cordage; juice of leaves and roots has been used medicinally in many ways.

Snakeplant—*Sansevieria trifasciata*

Sapindus saponaria Linn.   SAPINDACEAE
SOAPBERRY—Native to South Florida, West Indies and northern South America.
Tree, to 30'; leaves pinnate with leaflets to 4" long; leafstems sometimes flanged with
oblong leaf-like wings; flowers in loose cluster up to 10" long; fruit round, to ¾"
across, henna-brown, contains saponin, has been used as a soap substitute and for
poisoning fish.

Saritaea magnifica Dugand (*Arrabidaea magnifica* Sprague)   BIGNONIACEAE
SARITA—Native to Colombia. Vine, woody, twining and climbing to treetops;
leaves evergreen, opposite, compound, having 2 oval, leathery, glossy leaflets to 4"
long with sometimes a tendril between them; flowers in clusters, tubular, with 5 lobes
flaring to a width of 2½", rose-purple, most abundant in winter but occur several
times a year. Seeds not produced in Florida. Grown from cuttings or air-layers taken
in spring. A vine of the highest ornamental quality, attractive at all times, but still not
commonly grown.

Schefflera actinophylla Harms (*Brassaia actinophylla*   ARALIACEAE
    Endl.)
QUEENSLAND UMBRELLA TREE; OCTOPUS TREE—Native to Australia.
Tree, to 40', sometimes forked, with relatively few short branches at top and occasional aerial roots; if cut back when young will have multiple trunks; leaves
evergreen, compound, consisting of several oblong-oval leaflets up to 1' long on 3"
stems, arranged in a whorl at the tip of each 1½ to 2' leafstalk. Detached and held upright, a leafstalk, topped by the whorl of leaflets, suggests an umbrella. The flowers
are small, red and massed along the entire length of erect flowering branches that
stand out above the foliage and have been likened to the arms of an octopus. It is
rapid-growing and sometimes grows on and encompasses another tree as does the
Strangler Fig, and this trait further suggests an octopus. The fruits are small, round,
red. Propagated by seed, cuttings, or air-layers. Increasingly popular in South
Florida landscaping despite spreading surface root system.

Queensland Umbrella Tree—*Schefflera actinophylla*

Schinus terebinthifolius Raddi   ANACARDIACEAE
BRAZILIAN PEPPERTREE; erroneously "Florida Holly"—Native to Brazil. Tree, to 40', very bushy and spreading, its long branches densely clothed with foliage; leaves evergreen, pinnate; leaflets to 3" long, dark-green above, light beneath, pungently aromatic when crushed; flowers tiny, whitish, in small clusters massed at the ends of the branches; male and female on separate trees; fruit round, ⅛" across, red, abundant and showy. A common ornamental in South Florida where its fruiting sprays are used for Christmas decoration. It is fast-growing and seedlings planted by birds have formed extensive "jungles" crowding out native vegetation. When in bloom, the plant is a common cause of skin and respiratory irritation.

Sechium edule Sw.   CUCURBITACEAE
CHAYOTE—Native to tropical America. Vine, herbaceous, with thick, hairy stems and large tendrils; developing a huge, starchy, edible root; leaves triangular, more or less lobed, to 8" long; flowers small, white, male and female clustered together; fruit pear-shaped, more or less ridged, smooth-skinned or with few or many prickles, puckered and indented at apex; skin light-green or whitish; flesh crisp, white, mild-flavored; seed flattened, fleshy, to 2" long. Fruit cooked as a vegetable. When over-mature, the apex yawns slightly, the seed protrudes and soon sprouts. The seed or entire fruit may be planted. Fast-growing; needs fence or arbor. Young leaves and shoots are edible and nutritious.

Selenicereus pteranthus Britt. & Rose   CACTACEAE
SNAKE CACTUS—Native to Mexico. Cactus, with 4- to 5-angled, ridged stems to 2" thick bearing numerous tufts of whitish hairs accompanied by 1 to 4 conic spines; creeping and climbing in serpentine tangles on walls and trees; flowers night-blooming, highly fragrant, to 1' long, with numerous white petals and stamens; fruit round, red, to 3" long. Abundant in coastal hammocks below Ft. Pierce where it was introduced during Seminole War years. *S. coniflorus* Britt. & Rose, also from Mexico, is naturalized inland in pinewoods.

Senecio confusus Britten   COMPOSITAE
MEXICAN FLAME VINE; ORANGE GLOW-VINE—Native to Mexico. Vine, woody, high-climbing; leaves evergreen, tongue-shaped, pointed, coarsely toothed, up to 4" long; flower-heads daisy-like, vivid orange, darkening to reddish-orange, up to 1½" across, in clusters. Blooms mainly in spring and summer. Grown from seed, cuttings or air-layers. Not common in South Florida. Becomes intertwined and tangled: pruning may result in skin rash.

Serenoa repens Small (*S. serrulata* Hook. f.)   PALMAE
SAW PALMETTO—Native from South Carolina to Florida Keys. Dwarf palm, with creeping stem to 8' long; sometimes grows erect to a height of 20' or more; leaves

Saw Palmetto—*Serenoa repens*

to 4' across, fan-shaped, deeply divided into narrow segments which are stiffly out-spread; stalks edged with sharp spines, or spineless; flowers small, white, densely massed in elongated, branched, plume-like clusters, fragrant; fruit black, oblong, to ¾" in length, eaten by the Indians and employed in medicine. Terminal bud edible; stems yield tannin, cork substitute and fiber for wallboard; leaves used to make small fans. Very common in pinelands and on sandy coasts of Florida; a major source of nectar for honeybees.

Sesbania grandiflora Pers. (*Agati grandiflora* Desv.) LEGUMINOSAE SESBAN; CORKWOOD TREE; PARROT FLOWER; VEGETABLE HUMMING-BIRD—Native to tropical Asia. Tree, to 40', but usually smaller, with thick, furrowed bark and soft wood; leaves pinnate, 6" to 1' long, with 20 to 60 oblong leaflets 1" in length; flowers pea-like, white, rose or maroon, 4" long; fruit a pod ½" wide and up to 2' long. Young fruits, foliage and flowers edible; mature seeds not eaten; bark medicinal. Raised from seed; fast-growing and short-lived. Occasionally found semi-wild on lower Keys.

Setcreasea purpurea Boom (*Tradescantia pallida* Hunt) COMMELINACEAE PURPLE QUEEN—Native to Mexico. Herb, succulent, with purple, reclining or climbing stems, and alternate, clasping, lance-shaped, pointed, purple leaves, and 3-petaled, lavender-pink flowers. Cuttings and slips take root readily. Introduced into South Florida as a landscape plant about 1953; soon became a common ground cover in private and commercial grounds and in parks and parkways. Stems are tender, break easily and the watery juice may irritate skin.

[138]

Severinia buxifolia Tenore (*Atalantia buxifolia* Oliv.)　　　　RUTACEAE
CHINESE BOX-ORANGE—Native to southern China. Shrub or small tree, thorny, to 6' tall; leaves evergreen, to 1½" long; flowers small, white, single or in small clusters; fruit round, about ⅜" across, black. Slow-growing from seed or cuttings; excellent as a low hedge or border for walkways in full sun. According to the late Dr. Walter T. Swingle, authority on *Citrus* and relatives, this plant "can withstand unusually large amounts of salt in the soil."

Sideroxylon foetidissimum Jacq. (*Mastichodendron foetidissimum* H. J. Lam)
　　　　　　　　　　　　　　　　　　　　　　　　　　SAPOTACEAE
MASTIC TREE; MASTIC JUNGLEPLUM—Native to South Florida, the Bahamas and West Indies. Tree, to 70', with shaggy bark; leaves evergreen, to 8" long, wavy-edged, clustered near ends of branches; flowers yellow, small; fruit (spring) yellow, oblong, up to ¾" long, shiny; pulp white, seed, oblong, brown. Fruit edible but somewhat bitter and contains sticky white latex.

Simarouba glauca DC.　　　　　　　　　　SIMAROUBACEAE
PARADISE TREE—Native to South Florida, the Bahamas and West Indies. Tree, to 50', with slender proportions; leaves evergreen, pinnate with oval leaflets up to 4" long, dark-green, glossy on top, light on underside; new growth bright-red; flowers cream-colored, small, profuse, in clusters to 1½' long; fruit (summer) oval, 1" long, changes from red to nearly black as it ripens; edible but insipid. Seed kernel yields culinary oil. Slow-growing from seed. Rarely planted.

Solandra nitida Zucc. (*S. maxima* P. S. Green)　　　　SOLANACEAE
CHALICE VINE; CUP-OF-GOLD—Native to Mexico. Climbing shrub, to 20', sometimes with aerial roots; leaves broadly oval, to 6" long, downy on underside; flowers single, to 9" in length, like long-stemmed goblets, cream-colored when first open, turning gradually to deep-yellow, fragrant at night; fruit round, to 2½" wide, whitish to pale-yellow, clasped loosely by long, green sepals, contains many small seeds. Commonest winter-blooming vine in South Florida; ingestion of flowers has caused serious poisoning.

Solanum seaforthianum Andr.　　　　　　　　SOLANACEAE
BRAZILIAN NIGHTSHADE—Considered probably native to Brazil. Vine, climbing or trailing, slender-stemmed; leaves to 8" long with lobes or 5 to 9 leaflets to 2" long, sometimes wavy-margined; flowers star-like, blue or lavender, up to 1" across, in hanging clusters; fruit almost round, ⅜" wide, bright-red, very attractive to birds; the unripe fruits probably toxic to humans. A graceful vine for limited space.

Chinese Box Orange—*Severinia buxifolia*　　　　Chalice Vine—*Solandra nitida*

Solanum wendlandii Hook.   SOLANACEAE
MARRIAGE VINE; DIVORCE VINE; COSTA RICAN NIGHTSHADE—Native
to Costa Rica. Vine, climbing, with stout, twining stems, few thorns; leaves are of at
least two types, simple or deeply 3-lobed, the latter to 10" long; some prickles on
midrib and leafstems; flowers lavender-blue, up to 2½" wide in showy clusters to 1'
across; fruit round, 2" or more wide. A rampant grower, from cuttings or air-layers.

Spathiphyllum spp.   ARACEAE
Native to tropical America and Malaysia. Herbs, peren-
nial, with several clustered, short- or long-stemmed,
more or less elliptic, pointed, glossy, erect, arching leaves
to 1½' long; flowerstalk, rising above the foliage, bears
at the tip a showy, upward-pointing white or greenish
bract and a short, knobby spike of tiny flowers. Most
common in patios is a hybrid of *S. kochii* Engl. & Krause
popularly called "*S. clevelandii*". Grown from seed or by
division, in shade.

Spathodea campanulata Beauv.   BIGNONIACEAE
AFRICAN TULIP TREE; FOUNTAIN TREE—Native to tropical Africa. Tree, to
70'; leaves semi-evergreen, compound with oval leaflets to 5" long; flowers (winter)
tulip-like, 2" wide, orange-scarlet, in striking clusters extending up above the foliage.
Buds hold water and African children use them in play as water pistols. Seed pod,
rarely produced, oblong, to 9" in length and 1½" wide, pointed at both ends, brown
when mature, contains white-winged seeds. Leaves, flowers and bark used medicinally; seed edible. Fast-growing from seed and cuttings. Has been a popular ornamental
in South Florida but many have succumbed to hurricanes.

Spondias purpurea Linn.   ANACARDIACEAE
PURPLE MOMBIN; SCARLET PLUM; SPANISH PLUM—Native to tropical
America. Tree, to 25', open-branched, straggly; leaves pinnate in attractive clusters
with leaflets up to 1½" long; tree bare during winter; small rose-purple flowers
appear in spring before the leaves, as do the fruits. Fruits up to 1½" long, plum-like,
broad-oval, sometimes with a knobby, blunt tip; skin smooth, tough, red to dark-
purple, sometimes yellow-striped; pulp yellowish, musky, juicy; seed large, oval or
knobbed, encased in fiber. A form locally called Hog Plum, bears in the fall and has
clear yellow fruits with fairly smooth, oblong seeds. Propagated by cuttings.

Stapelia gigantea N. E. Br.   ASCLEPIADACEAE
CARRION FLOWER—Native to tropical and southern Africa. Herb, succulent,
cactus-like, with erect, slightly curved, 4-angled, toothed stems to 8" high and 1¼"
thick, forming colonies; leaves rarely produced; flowers star-shaped, to 15" wide,
buff-colored with fine brown crossbars and a coating of purple hairs, evil-smelling.
Propagated by division; thrives and spreads in full sun and dry soil; blooms more
than once a year.

Stephanotis floribunda Brongn. ASCLEPIADACEAE
BRIDAL BOUQUET; MADAGASCAR
JASMINE—Native to Madagascar. Vine, with woody, twining stems, to 15'; leaves evergreen, oval, smooth-edged, up to 4" long, leathery, glossy; flowers (June-July) to 2" long, tubular with 5 flaring lobes, waxy, white, in clusters of 5 to 8, fragrant; fruit a horn-like seed pod up to 4" long. Flowers favored for wedding bouquets. Grown from seed or cuttings in sun or partial shade, preferably where its roots can go under a wall as it is susceptible to nematodes. An elegant vine that should be more commonly cultivated.

Sterculia foetida Linn. STERCULIACEAE
BANGAR NUT—Native from India to Australia and East Africa. Tree, to 60 or 100' with spreading, horizontal branches; leaves, clustered at the branch tips, deciduous, digitately compound with 5 to 9 elliptic, pointed leaflets 1 to 1½' long; flowers, in drooping sprays (Feb.-March), 1½" wide, petalless with 5 red-and-yellow or purple, velvety, recurved sepals, emitting a far-reaching, foul stench; fruit a 2-lobed, woody, dark-red pod, 4" wide, which splits open and reveals 10 to 15 black, oblong seeds, ¾ to 1" long, clinging like teeth to the rim of each half. Seeds toxic raw; have been roasted and eaten but even so are not wholesome. Fast-growing from seed.

Strelitzia reginae Banks STRELITZIACEAE
BIRD-OF-PARADISE FLOWER; CRANE'S BILL—Native to South Africa. Herb, to 4', with erect, oblong leaves to 1½' long and 6" wide on long stems; inflorescence has fancied resemblance to a bird with raised wings, the conspicuous features being a nearly horizontal, boat-shaped, purple- or red-tinted spathe about 6" long, and the half-dozen, 3- to 4" blue and orange-yellow flowers extending obliquely back from it; seeds edible. A much-prized plant of florists and subtropical gardens. Grown from seed or offshoots.

Strumpfia maritima Jacq. RUBIACEAE
STRUMPFIA—Native to Florida Keys, the Bahamas and West Indies. Shrub, to 10', bushy, downy; leaves needle-like, to 1" long in 3's, woolly-white on underside; flowers white or pink, 3/16" across; fruit round, to ¼" across, white or red. In Curacao, the bush forms dense masses dangling from crevices high on the face of rocky cliffs.

Suriana maritima Linn. SURIANACEAE
BAY CEDAR; TASSEL PLANT; THATCH LEAF—Native to South Florida, the Bahamas, West Indies, tropical America and Old World tropics. Shrub or tree, to 15', bushy; leaves paddle-shaped, fleshy, woolly, to 2" long, in clusters at ends of twigs; flowers yellow, 5-petaled, ½" wide; fruit composed of 4 nestled, round, hairy, nut-like parts, each 3/16" wide. Grows on beaches or rocky shores. Seedlings just a few inches high can be transplanted for coastal landscaping.

Swietenia mahagoni Jacq.   MELIACEAE
MAHOGANY—Native to South Florida, the Bahamas, West Indies and Central and South America. Tree, to 75' with compact, rounded head; leaves briefly deciduous, pinnate with pointed leaflets to 4" long, dark-green on top, yellowish or brownish on underside; flowers greenish-yellow, tiny, in small clusters; fruit a woody, light-brown, conical and erect pod to 5" long and 3½" broad, splits and opens like an umbrella to release the winged seeds. Wood dark reddish-brown, valued for furniture. Fast-growing from seed. Often planted as a street and shade tree in South Florida.

Syngonium podophyllum Schott   ARACEAE
SYNGONIUM—Native from southern Mexico to Panama. Vine, succulent, epiphytic, with aerial roots; leaves variable, arrowhead-shaped when young (when it is called *Nepthytis* by nurserymen); older leaves are divided into 3 to 5 elliptic, pointed leaflets; mature leaves may have 7 to 11 leaflets in a whorl; leaves may be entirely green or attractively variegated with pale grayish-green. Rarely blooms. Grows fast from cuttings. Common in patios and climbing cabbage or date palms; important in Florida's commercial foliage plant industry, especially potted with "totem pole" support.

Synsepalum dulcificum Daniell   SAPOTACEAE
MIRACLE FRUIT—Native to West Tropical Africa. Shrub, or small tree, to 6' or more; leaves clustered near branch tips, evergreen, obovate, to 4" long; flowers (spring and fall) inconspicuous, brownish, in small clusters; fruit oblong, pointed, ¾" long, shiny, red, with 1 large seed. Fruit, thoroughly chewed, acts as a taste converter, for a sour strawberry or lime juice put into the mouth 15 minutes to 3 hours later will taste delightfully sweet. Slow-growing from seed. Fruits from Florida plants grown as curiosities for many years, have, since 1962, been supplied to various research workers endeavoring to isolate the modifying property for dietetic purposes.

Syzygium cumini Skeels (*Eugenia jambolana* Lam.)   MYRTACEAE
JAMBOLAN—Native to India, Burma, Ceylon and the Andaman Islands. Tree, to 50 or even 100', with broad crown; leaves evergreen, opposite, elliptic, to 4" wide and 10" long; flowers white, turning pink, funnelform, 1" long, in small clusters; fruit (summer) round or oblong, to 2" long; skin, smooth, turning from green to red and finally purple-black; pulp white or purple, juicy, with 1 seed to 1½" long. The fruit, acid to sweet and somewhat astringent may be used for jelly, juice or wine. Fast-growing from seed, becomes too large for the average home yard and bears so heavily that the fallen fruits present a major disposal problem.

Syzygium jambos Alston (*Eugenia jambos* Linn.)   MYRTACEAE
ROSE APPLE—Native to East Indies. Tree, to 30' with spreading top and supple, drooping branches; leaves evergreen, slender, pointed, to 8" long; new growth pink;

flowers whitish, consisting mainly of a brush-like tuft of stamens, up to 2½" across; fruit nearly round, to 2" wide, pale-yellow with calyx protruding at apex; rose-like in aroma; flesh crisp; center hollow, containing one or two rough, brown seeds. Fruit has a sweet, rose-like flavor and is eaten raw or preserved. Grown from seeds. Found in fruit tree collections and on some of the older homesteads in South Florida.

Tabebuia argentea Britt.   BIGNONIACEAE
SILVER TRUMPET-TREE; GOLDEN BELL—Native to Paraguay. Tree, to 25', with crooked trunk and compact head, rough, corrugated, light-gray bark; leaves deciduous, divided into 5 to 7 somewhat stiff, narrow leaflets to 6" long, silvery, clustered at ends of branches; flowers (spring) yellow, bell-shaped, up to 2½" long, in profuse clusters when tree is bare; fruit a gray, black-streaked pod up to 4" long. Propagated by seed and grafting; grows at moderate rate. A glorious flowering tree, requiring little space.

Tabebuia palmeri Rose   BIGNONIACEAE
Native to southwestern Mexico. Tree, to 60'; leaves deciduous, compound, with 5 elliptic, pointed leaflets to 5½" long, more or less downy; flowers (December-March), in terminal clusters, funnelform, to 3" long, pink or nearly violet; seed pod smooth, 1' long and ⅝" wide. Selections propagated by grafting. Charming hybrids between this and other species have been distributed by local growers. A *"T. pallida"* and *T. haemantha* hybrid with large reddish flowers is sold in nurseries as "Carib Queen".

Tabebuia rosea DC. (*T. pentaphylla* auth. NOT Hemsl.)   BIGNONIACEAE
PINK TRUMPET-TREE—Native to West Indies, Central America, northern South America. Tree, to 60', erect, of medium proportions, with rough, gray bark; leaves compound with 5 to 7 stiff, 6"-leaflets on longish stems and arranged in a whorl at the tip of the leafstem; flowers (spring) pink, like slender bells, to 3" long, in clusters; fruit a slender pod up to 1' in length. Types with smaller leaves (some evergreen, some deciduous) and paler flowers have been much planted in recent years as door-yard and parkway trees under the name *"T. pallida* Miers" regarded by some as a synonym of *T. rosea*. Some trees bloom twice a year. These are fine ornamentals, re-quiring little space and deserving of popularity.

Tamarindus indica Linn.   LEGUMINOSAE
TAMARIND—Native to East Indies. Tree, to 80', with enormous, domed top when old; foliage evergreen, feathery, dense; flowers yellow-and-red, to 1" wide, in small clusters; fruit a pod, usually 4 to 6" in length and to 1" wide. When young, the fruit is tender, green, very acid and used for seasoning; when mature, has brittle brown shell and contains several large seeds embedded in brown, tart pulp used in ade and for sweetmeats. Pulp is preserved by layering with sugar. Propagated by seed or air-layers; slow-growing, ornamental and impervious to windstorms.

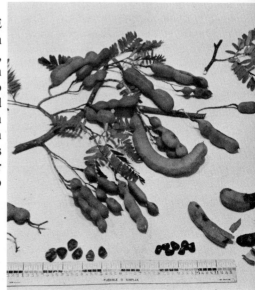

Taxodium distichum L.C. Rich   TAXODIACEAE
BALD CYPRESS—Native to swamps and lake shores from southern Delaware to
southern Florida, west to Texas and north to Oklahoma. Tree, to 150', strongly but-
tressed at base, and producing from its spreading roots, vertical, woody peaks (to 2'
high) called "knees"; leaves deciduous, alternate, linear, ½ to ¾ " long, on deciduous
twigs; flowers tiny, male in long strings, female in short clusters; cone (winter) nearly
round, to 1" wide, turning brown and releasing brown seeds ½" long. Native stands
much reduced by logging. Fairly fast-growing from seed and is successfully grown as
an ornamental on high land; is particularly beautiful when fresh new foliage appears
in spring. Polished "knees" sold as novelties and used for lamp bases.

Taxodium distichum var. *nutans* Sweet (*T. ascendens* Brongn.)   TAXODIACEAE
POND CYPRESS—Native to shallow ponds and moist grounds in the coastal plain
from southeast Virginia to southern Florida and Louisiana. Tree, to 75', with but-
tressed or swollen base and few "knees"; leaves deciduous, alternate, 1/3" long, upturned and pressed close to the string-like branchlets which are also deciduous; cone
(winter) rounded, 1" wide, dangling from bare branches, contains several brown,
½"-long seeds. Grows from seed more slowly than the bald cypress, and is useful for
highway and home landscaping even in well-drained limestone areas.

Tecoma stans HBK (*Stenolobium stans* Seem.)   BIGNONIACEAE
YELLOW ELDER; PLOPPER: YELLOW BIGNONIA—Native to West Indies
and northern South America; naturalized in Florida. Shrub or small tree, to 25',
bushy; leaves deciduous, pinnate, with slender, light-green leaflets to 5" long; flowers
bright-yellow, bell-like, up to 2" long, in nodding, showy clusters; fruit a slim pod to
8" long. Springs up like a weed, forming small thickets on neglected land. The tree is
so lovely for 2 or 3 weeks in the fall that it should be planted wherever possible even
though it is shaggy with dry, split seed pods during the winter. Wild trees transplanted
and cut back make quick regrowth.

Tecomaria capensis Spach.   BIGNONIACEAE
CAPE HONEYSUCKLE; HAPPY VINE—Native to South Africa. Shrub, semi-
climbing, with slender branches to 6' in length; leaves evergreen, attractive, pinnate
with pointed, toothed leaflets to 2" long; flowers (all year) up to 2" long, slender
funnel-shaped, slightly flaring into 4 segments, reddish-orange or scarlet, in
elongated clusters, visited by humming-birds; fruit a pod up to 6" or more in length.
Trailing stems and cuttings take root readily. A versatile plant, useful as a clipped
shrub, hedge or vine.

Terminalia arjuna Wight & Arn.   COMBRETACEAE
ARJAN—Native to India and Ceylon. Tree, to 80', with open, spreading top and
fairly smooth, pinkish-brown bark; leaves deciduous, oblong, to 8" long and 2" wide,
with 2 round glands at base of leafblade; flowers yellow; fruit a capsule, 1" long, with
5 or 6 prominent vertical wings. Wood hard, used for boats and in building; bark
used in tanning and yields medicinal products. A fast-growing shade tree, raised from
seed.

[144]

Bald Cypress—*Taxodium distichum*   Tropical Almond—*Terminalia catappa*

Terminalia catappa Linn.   COMBRETACEAE
TROPICAL ALMOND; INDIA ALMOND—Native to Madagascar, Malaya, East
Indies. Tree, to 80', with widely separated tiers of nearly horizontal branches; leaves
paddle-shaped, leathery, to 1' long and 6" broad, clustered at ends of twigs, turn red
and fall in winter but are quickly replaced by glossy new growth; flowers whitish,
tiny, in "rat-tail" spikes; fruit oval, to 2¼" long, yellow-green with pink blush, thin
layer of flesh and large fibrous seed enclosing slender kernel. Flesh and kernel edible;
fruit yields dye and ink and is used medicinally, as are leaves, bark and roots. Grows
quickly from seed; wind- and salt-resistant. Fairly common ornamental in South
Florida.

Terminalia muelleri Benth.   COMBRETACEAE
MUELLER TERMINALIA—Native to Australia. Tree, small to medium, with
spreading top; bark rough, dark-gray; curved twigs stand upright on upper surface of
the horizontal branches; leaves evergreen, clustered at ends of twigs, broad-oval, up
to 4" long, somewhat leathery, veins prominent on underside; flowers white, small, in
loose spikes; fruit dark-blue, ovoid, up to ¾" long, acid. Wood strong, used for axe-
handles. Fast-growing from seed.

Tetrapanax papyriferum Koch.   ARALIACEAE
RICE-PAPER PLANT—Native to Formosa and South China. Shrub, or small tree,
to 30', with trunk 2 to 4" thick; leaves evergreen, to 2' wide on 2½' stems; nearly circular, with 7 or 8 main lobes, deeply grooved and subdivided into smaller lobes,
toothed and wavy-edged, dull, dark-green above, downy white beneath; flowers
(winter), white, in fuzzy, globose heads borne in several great plumes from the top of
the stem. Handsome ornamental, fast-growing from seed, but suckers freely and may
become a nuisance. Stem pith of young plants made into rice paper.

Rice Paper Plant—*Tetrapanax papyriferum*

Florida Tetrazygia—*Tetrazygia bicolor*

Tetrazygia bicolor Cogn.   MELASTOMACEAE
FLORIDA TETRAZYGIA—Native to southern Florida, the Keys, the Bahamas
and Cuba. Shrub, or small tree, to 30'; leaves evergreen, opposite, glossy, dark-green
above, pale beneath, lance-shaped, to 5" long, with 3 conspicuous longitudinal veins
and faint horizontal veins; flower (spring-summer) white with yellow center, 4- or 5-
petaled, in plump, erect clusters to 5" long; fruit purple-black, 1/3" long, round with
protruding calyx at apex. One of the loveliest of our native shrubs, abundant in
pinelands and hammocks and worthy of cultivation.

Thespesia populnea Soland.   MALVACEAE
SEASIDE MAHOE; CORK TREE; PORTIA TREE;
FALSE ROSEWOOD—Native to tropical coasts of Old
World; naturalized in South Florida and on the Keys.
Tree, to 50', with spreading branches; leaves evergreen,
heart-shaped, pointed, to 4" long; flowers to 3" wide,
hibiscus-like, light-yellow in morning with purple-red
center, petals turn dark-red in afternoon; flowers remain
on tree for several days; fruit a woody, 5-parted capsule
up to 1½" across. Wood of tree is multi-colored with
shades of red and purple, rose-scented when freshly cut.
Juice of immature fruit and bark medicinal; seeds yield
oil; bark yields red dye. Grown from seed, cuttings or
air-layers as a street tree; salt-tolerant.

Thevetia peruviana Schum. (*T. nereifolia* Juss.)   APOCYNACEAE
LUCKY NUT; YELLOW OLEANDER; BE-STILL TREE—Native to tropical
America. Shrub or tree, to 30', bushy; leaves very narrow, up to 7" long, bright-green,
glossy; flowers bell-like, to 3" long, yellow or peach-colored, fragrant; fruit an angled
capsule, 1½" wide, green at first, turning black and dry and containing a somewhat
diamond-shaped brown stone enclosing flat seeds that are highly toxic, as are the
bark and other parts of the plant. The polished stones are used for novelties and
carried in the pocket to bring good luck or afford imagined protection from ills. Occasionally planted as an ornamental in South Florida but often defoliated by cater-
pillars in winter.

Thrinax radiata Lodd (*T. floridana* Sarg.;   PALMAE
   T. parviflora auct. non Sw.; *T. wendlandiana* Becc.)
FLORIDA THATCH PALM: JAMAICA THATCH PALM; SILK-TOP
THATCH PALM—Native to South Florida, the Keys, Bahamas and West Indies.
Palm tree, to 30'; trunk 6" thick, rough until old; leaves to 3' long, fan-like, almost
circular and not deeply divided, yellowish-green, not silvery on underside; flowers
small, whitish, in clusters to 3½' long: fruit round, ivory, ½" across.

Thrinax morrisii H. Wendl. (*T. microcarpa* Sarg.; *T. keyensis* Sarg.)   PALMAE
BRITTLE THATCH PALM; KEY THATCH PALM—Native to Florida Keys,
Bahamas and Cuba. Palm tree, to 35'; leaves fan-like, light-green on top, silvery or
bluish on the underside, up to 4' wide on 4' stalks; fruit round, white, 1/3" across.
Seeds remain viable 2 to 3 months.

Thuja orientalis Linn.   CUPRESSACEAE
ORIENTAL ARBORVITAE; WHITE CEDAR—Native to northern China and
Korea. Tree, to 40', dense, pyramidal or rounded, evergreen, with reddish bark; short
branches and slender branchlets, clothed with tiny, scale-like, rich-green leaves,
resemble coarse ferns, stand upright in compact mass; cone up to 1" long, spiny. A
handsome ornamental with many cultivated forms, some having bluish or yellowish
foliage. Fast-growing from cuttings; small specimens often mistakenly planted
beneath windows which they soon block.

Thunbergia erecta T. Anders.   THUNBERGIACEAE
                                              (formerly ACANTHACEAE)
BUSH CLOCKVINE; KING'S MANTLE—Native to tropical Africa. Shrub, to 8',
bushy, sprawling; leaves evergreen, to 3" long, broad-oval at base, angled to a point
at the tip, sometimes toothed; flowers single, bell-like, 5-lobed, flaring, up to 3" long,
blue or purple with yellow throat. Base of flower clasped by two green, leaf-like
bracts. Variety *alba* has white flowers. Everblooming in sun; should be more commonly planted.

Thunbergia fragrans Roxb.   THUNBERGIACEAE
SWEET CLOCKVINE—Native to India. Vine, with slender, rather woody, twining
stems; leaves evergreen, irregularly oval, pointed, rough, to 3" long; flowers non-
fragrant, borne singly, up to 2" wide, with squared, centrally-pointed lobes, white,
base clasped by two green, leaf-like bracts; fruit a seed capsule to ¾" long, beaked.
Fast-growing from seeds; dainty and attractive on a trellis but must be kept under
control; tends to run wild.

Thunbergia grandiflora Roxb.   THUNBERGIACEAE
SKY-VINE; BENGAL CLOCKVINE—Native to Bengal. Vine, woody, twining;
leaves evergreen, to 8" long, irregularly oval, long-pointed and somewhat lobed or
toothed; flowers (winter-spring) up to 3" wide, 5-lobed, sky-blue or dark-blue with a
whitish or pale-yellow throat, or, in variety *alba,* all white; single or in hanging
clusters. Base of flower clasped by two green, leaf-like bracts. Grown from cuttings;
vigorous; provides lush drapery for walls and fences.

Tibouchina Urvilleana Cogn.   MELASTOMACEAE
GLORYBUSH—Native to southern Brazil. Shrub, to 15', with quadrangular, hairy twigs; leaves evergreen, opposite, ovate, pointed, with 5 conspicuous, longitudinal veins, dark-green and downy above, silvery-hairy beneath; flowers (all summer) in clusters at the branch tips, 5-petaled, to 3" wide, rich purple. Grown from cuttings or air-layers in acid soil. In limestone areas, large holes should be prepared with sand and peat moss. Superb ornamental, long cultivated in Central Florida.

Tillandsia fasciculata Swartz   BROMELIACEAE
QUILL-LEAF TILLANDSIA; CARDINAL AIR-PLANT—Native to South Florida, the Bahamas, West Indies and Central America. Epiphytic herb, to 2'; leaves arching, 1½ to 3' long, tapering to a slender point, bluish; flowers blue, with green-and-red or ivory bracts arranged in a spike on erect, red stalk. Conspicuous on cypress and pine trees in early spring and summer.

Tillandsia flexuosa Sw. (*T. aloifolia* Hook.)   BROMELIACEAE
CORKSCREW AIR PLANT—Native to South Florida, the Bahamas, West Indies and tropical America. Epiphytic herb; leaves 1' long, twisted and arranged in spiral rosette, horizontally striped with alternating bands of light- and dark-green; flower stalk to 3½' tall, branched and bearing clusters of white, pink or purple flowers in late summer.

Tillandsia recurvata Linn.   BROMELIACEAE
BALL MOSS, or THREAD-LEAVED WILD PINE—Native from Florida to Texas, the Bahamas, and south to Argentina and Chile. Herb, with slim roots, epiphytic, entirely covered with silvery-gray scales; stems to 4" long; leaves linear, to 7" long, erect to extremely recurved; flowers few, blue, in slender erect spike. Very conspicuous plant, frequently in dense masses on branches of trees and shrubs and even on telephone and power lines.

Tillandsia usneoides Linn.   BROMELIACEAE
SPANISH MOSS, or TREEBEARD—Native from Virginia to Texas and south to Argentina and Chile. Herb, epiphytic, rootless, with threadlike, branching stems to 6' long and recurved leaves, 1 to 2" long, wholly coated with moisture-absorbing, minute, silvery scales. Flowers (summer) yellow-green or bluish, to ½" long; seed pod cylindrical, 1" long, containing tiny seeds bearing barbed hairs. Hangs in filmy tresses from trees, especially oaks, cypresses and pines. Not as abundant in South Florida as further north where it is gathered and the fiber used as upholstery and mattress filling.

Torrubia longifolia Britt. (*Guapira longifolia* Little)   NYCTAGINACEAE
BEEFWOOD; NARROW-LEAVED BLOLLY—Native to South Florida, the Bahamas and West Indies. Shrub or tree, to 50', with light bark; leaves slender, to 2½" long; flowers green or purplish, small, clustered; fruit oval, to ¼" long, red, ribbed, in dense, showy clusters.

Manila Palm—*Veitchia merrillii*

Zephyr Lily—*Zephyranthes* sp.

Tournefortia gnaphalodes R. Br.   BORAGINACEAE
BAY LAVENDER; SEA LAVENDER—Native to South Florida, the Bahamas,
West Indies, Mexico. Shrub, to 6' tall, growing in clumps; leaves narrow, whitish,
velvety, up to 4" long in terminal rosettes; flowers white, small, in clustered, one-
sided, curved spikes; fruit nearly round, to 3/16" wide, black. Forms handsome
clumps on the seashore. Propagated by seed or ground-layering. Elegant plant for
coastal landscaping.

Trevesia palmata Viz.   ARALIACEAE
Native to Southeast Asia and East Indies. Shrub, or
small tree, to 20', with few or no branches; stem and
branches hairy and prickly; leaves evergreen, leathery, to
2' wide, nearly circular, deeply-divided to a solid center,
each division deeply lobed and toothed. A shade plant
introduced into the United States by Mulford Foster of
Orlando and sold in Florida nurseries since 1954. *T.
palmata micholitzii* is speckled with silver and called
"Snowflake" in the trade. Grown from cuttings.

Tribulus cistoides Linn.   ZYGOPHYLLACEAE
PUNCTURE WEED; LARGE YELLOW CALTROP;
BILLY-GOAT WEED—Native to Georgia, Florida,
the southern Bahamas, West Indies and Central
America. Herb, perennial, with many stems growing
close to the ground, and extending to 3½' or more;
leaves pinnate with slender, oblong leaflets ½" long,
satiny on underside; flowers bright-yellow, 5-petaled, up
to 1½" across, begin to close at noon; fruit a 5-angled
capsule, spiny and painful to step on unshod. Grown as a
ground cover in sunny locations; often in coastal
parkways.

Trimeza martinicensis Herb.   IRIDACEAE
WALKING IRIS; ROOSTER'S TAIL—Native to
Jamaica, the Lesser Antilles and northern South
America. Herb, iris-like; leaves rise from the ground in
groups, sheathing one another at the base, sword-like,
tapering to a point, 2 to 3' long and up to 1½" wide;
flower stalk 1 to 3' high, bearing a succession of flowers
from a green, inch-long spathe; flowers 1½ to 2" across
with 6 petal segments, the outer three larger and longer
than the inner three, the latter with recurved tips; bright-
yellow with dark-brown or dark-red dots at base of
petals; close before end of day. New plants are produced
on the flower stalk and, as they form, the stalk bends
over until it touches the ground, allowing the plantlets to
take root, hence the name "walking iris." Very useful in
shaded locations and will also grow well in partial sun.

Yellow Alder—*Turnera ulmifolia*

Triphasia trifolia P. Wils.   RUTACEAE
LIMEBERRY—Probably native to southern Asia and East Indies. Shrub or tree, to
15', thorny; leaves evergreen, compound; leaflets up to 2" long, in 3's, dark-green;
flowers white, fragrant, ½" wide; fruit round-oval, to ⅝" long, maroon, glossy, with
scant sirupy pulp, large green seed. Fruit resinous, edible, made into a preserve in
China. Grown from seed or cuttings; makes a very attractive, compact, low hedge.

Turnera ulmifolia Linn.   TURNERACEAE
YELLOW ALDER—Native to Mexico, the Bahamas, West Indies and northern
South America. Shrub, slender, scarcely woody, to 4'; leaves evergreen, variable, up
to 4" long, sometimes downy-white on underside; flowers 5-petaled, 1 to 2" across,
yellow, fragrant, close at noon; fruit a seed pod about ¼" long. Variety *elegans* Hort.
has purple blotches at the base of the petals. Grown from seed in shade.

Vanilla planifolia Andr. (*V. fragrans* Ames)   ORCHIDACEAE
MEXICAN VANILLA—Native to Mexico and Central America. Climbing orchid,
with thick, succulent, green stems sending out aerial roots; leaves fleshy, leathery, to
9" long and 2" wide; flowers 2" long, 5-parted, and with one petal forming a lip in the
center; greenish-yellow, in long clusters; fruit a slender pod to 9" long, black when
cured and covered with minute crystals; used for flavoring and yields vanilla extract.
Rare and survives in South Florida only in a slathouse or greenhouse.

Veitchia merrillii H. E. Moore (*Adonidia merrillii* Becc.)   PALMAE
MANILA PALM; MERRILL PALM; sometimes sold as "Dwarf Royal
Palm"—Native to Philippine Islands. Palm tree to 25', trunk slender, tapering at top,
ringed; 9-12 leaves, feather-shaped, to 6' long, stiffly arched; leaflets up to 2½' long
and 2" wide; flowers small, in branched clusters below the crownshaft; fruit oval,
pointed, up to 1½" long, bright-red, highly ornamental. Bears mostly in midwinter
and midsummer; one cluster may be loaded with ripe fruits while other bunches are
still green. A favorite palm for dooryard planting in South Florida; grown from seed
less than 3 weeks old. Many have succumbed to the lethal yellowing disease which attacks the coconut palm.

[150]

Vitex angus-castus Linn.   VERBENACEAE
CHASTE TREE; HEMP TREE; MONK'S PEPPER TREE—Native to southern
Europe. Shrub or tree, to 10'; leaves deciduous, divided into 5 to 7 narrow, lance-
shaped, pointed leaflets up to 4" long, sometimes tooth-edged, dark-green on top,
downy-gray on underside, pungently aromatic; flowers lavender, small, in spikes to
7" long; spikes sometimes in clusters; fruit small, with 4 stones enclosing pepper-like
seeds used in seasoning food. Leaves and bark have medicinal uses; infusion of leaves
said to allay fever. Variety *alba* has white flowers. Propagated by cuttings.

Vitex trifolia Linn.   VERBENACEAE
Native to Asia, East Indies and Australia. Shrub, to 12' or sprawling; leaves
evergreen; leaflets in 3's or single, oblong, pointed, up to 3" long, dull-green above,
downy and whitish on underside; flowers ½" wide, lavender or blue with white dot,
in clusters up to 4" long. Variety *variegata* has white-variegated foliage. Leaves are
used medicinally, and are burned to repel insects. Grown from cuttings, usually as a
hedge, but expands rapidly and needs heavy pruning.

Vitis spp.   VITIDACEAE
GRAPES—Though commercial viticulture has never succeeded in South Florida, a
number of grape fanciers have raised flourishing grafted vines in home gardens and
plant breeders are still hopefully working to develop suitable hybrids by crossing
native species with introduced varieties. The so-called "Key Grape," of doubtful
origin, does well on the Keys and produces large clusters of fruits, small, but of fair
quality. There are several wild species which produce tart fruits.

Washingtonia robusta H. Wendl.   PALMAE
   (*W. gracilis* Parish)

Mexican Washington Palm
Washingtonia robusta

MEXICAN WASHINGTON PALM—Native to
northern Mexico, including Lower California. Palm tree,
to 80', with slender trunk; leaves fan-like, to 3' across, on
3', reddish-brown stalks which are spiny-edged especially when young; old leaves hang down around trunk,
forming a "beard" or "skirt" which is often trimmed
short; flowers small, white, in branched clusters; fruit
round, black. Terminal bud and fruit edible and seeds
made into meal; leaves used in making mats and baskets. Very fast-growing from seed. Bird-planted
seedlings spring up in cultivated fields and vacant lots.

Wedelia trilobata Hitchc.   COMPOSITAE
TRAILING WEDELIA—Native to West Indies, Central and northern South America; naturalized in
Florida. Herb, with trailing stems which may climb or
extend 3' or more and take root; short upright branches;
leaves slightly fleshy, opposite, variable, sometimes 3-
lobed, toothed, up to 3 or 4" long; flower-head yellow,

Trailing Wedelia
Wedelia trilobata

10-rayed, each ray 3-toothed at tip. A vigorous ground cover for both shady and sunny areas; needs cutting back occasionally; clippings used to establish new beds.

Xanthosoma spp.   ARACEAE

COCOYAM; ELEPHANT'S EAR; MALANGA, YAUTIA, or TANIER—Native to the West Indies and northeastern South America. Herbs, to 8' high, with starchy rhizomes, sending up a clump of long-petioled leaves, the blades arrowhead-like, to 3½' long, attached to the petiole at the cleft; flowers small, white, in thick spadix to 5" long enclosed in ivory spathe. *X. caracu* Koch & Bouche is a very common dooryard plant, now grown in fields in South Florida to supply the demand for the rhizomes which are staples in the diet of many Cuban and Puerto Rican people living in the United States. The plant and rhizomes contain irritating principles which affect the mouth and throat if bitten into when raw. The GIANT COCOYAM (*X. jacquinii* Schott.) is especially acrid, but a striking ornamental for shady locations and blooms more freely than other species. *X. nigrum* Mansf. (formerly *X. violaceum* Schott.) has purple petioles and the leaf blade is often tinged with purple.

Ximenia americana Linn.   OLACACEAE

TALLOWWOOD PLUM; MOUNTAIN PLUM; PURGE NUT—Native to the tropics of both hemispheres, including South Florida. Tree, to 25', spiny; leaves evergreen, to 3" long; flowers yellow, ⅜" wide, fuzzy, in small clusters; fruit plumlike, oval, up to 1½" long, bright-yellow, smooth, juicy, acid, almond-flavored; seed oval, white, with nut-like kernel which is edible but possibly purgative and yields oil for food use, soap and lubricating; fruit eaten raw and made into beer; bark of tree used in tanning; various parts of plant used medicinally.

Spanish Bayonet—*Yucca aloifolia*

Spineless Yucca—*Yucca elephantipes*

Yucca aloifolia Linn.   AGAVACEAE
SPANISH BAYONET; ALOE YUCCA—Native to southern U.S. and West Indies.
Herb, with thick, erect stem to 8', even 20', high, densely covered with stiff,
"bayonet" leaves to 2 ½' long and 2 ½" wide, tapering, pointed and sharply needle-
tipped; flowers white, fleshy, cup-like, to 4" across, in erect cluster up to 1 ½' tall;
fruit nearly cylindrical, up to 5" long, purple-skinned. Flowers edible when fresh;
fruit edible but bitter and seedy; fiber of leaves used for cordage. Propagated by
suckers. Unwisely planted in foundation boxes, planters and near walkways. Should
be isolated or spines clipped from leaftips. Salt- and wind-resistant, ideal for coastal
planting.

Yucca elephantipes Regel   AGAVACEAE
SPINELESS YUCCA—Native to Central America and southern Mexico. Tree, to
30', with thick stem, swollen at the base, and with a few short branches; leaves closely
set all around the upper stem and branches, to 3' long, dark-green, strap-like, to 3"
wide, stiff, smooth with slightly rough edges but without terminal spine; flowers
(spring-summer) bell-shaped, waxy, white, in tall, erect cluster. Fruit fleshy, oblong,
seedy. Grown from seed or cuttings in sun or shade on wet or dry land. Promoted as a
safe *Yucca* for Florida gardens since 1956, but inclined to grow too large for most
home landscaping purposes. Common tall hedge plant in Central America where the
flowers are a staple food.

[153]

Zamia floridana DC. and *Z. integrifolia* Ait.   CYCADACEAE
COONTIE; FLORIDA ARROWROOT; SEMINOLE BREAD—Native to Florida, including the Keys. Fern-like plants; leaves erect, stiff; feather-shaped, to 3' long and 8" or more wide, with narrow, glossy, dark-green leaflets and one to several cylindrical, velvety, reddish-brown cones, the male to 7" in length and slender, the female (on separate plant) shorter and broader. The female cone, on maturity, releases a mass of angled seeds, 1" long, with fleshy, bright-orange or scarlet seedcoats. Thick underground stem, after removal of poisonous principle by maceration and washing, yields starch which was much used by Seminoles and early settlers. Grown from seed or young plants transplanted from the wild; in sun or shade; drought-tolerant.

Zanthoxylum fagara Sarg.   RUTACEAE
LIME PRICKLY ASH; WILD LIME; SATINWOOD—Native to South Florida, the Keys, Bahamas, West Indies and tropical America. Tree, to 30', with slender branches, rough bark and hooked spines; leaves evergreen, pinnate with winged petiole and rachis and glossy leaflets up to 1¼" long; flowers tiny, greenish, in small clusters; fruit round, ⅛" across, brown. Foliage has lime aroma when bruised.

Zebrina pendula Schnizl.   COMMELINACEAE
WANDERING JEW—Native to the West Indies and Central America. Herb, succulent, with slender, trailing, branching stems, rooting at nodes; leaves ovate, pointed, to 2" long, upperside shining, purplish-green with 2 lengthwise stripes of silver, underside bright-purple; flowers, 3-lobed, red above, white below, flanked by leaf-like bracts. Universally grown in pots and baskets and a very useful ground cover, to 15" high, for shaded spots in South Florida. Grows readily from cuttings; apt to overrun allotted space.

Zephyranthes spp.   AMARYLLIDACEAE
ZEPHYR LILY; RAIN LILY—Herb, of dainty aspect, with small bulb and erect, grass-like leaves to 1½' tall; flowers 1 to 2" wide, with 6 upright petals flaring from a short tube; borne singly on upright hollow stems. *Z. atamasco* Herb., the ATAMASCO LILY, with white flowers, is native to the southeastern U.S., including Florida; *Z. grandiflora* Lindl., with pink or rose flowers, is native to the West Indies and tropical America. Propagated by offsets. Die down in dry seasons; spring up and bloom during rainy periods.

Zizyphus mauritiana Lam.   RHAMNACEAE
INDIAN JUJUBE—Native to southeastern Asia. Tree, to 40', with fairly open crown of thorny, weeping branches; leaves evergreen, alternate, round-oval, to 3" long, deep-green and shining above, downy-white or rusty on the underside; flowers whitish, ⅛" wide, very fragrant; fruit (January-February) oval, to 1" long, smooth, turns from green to yellow when it is crisp and crabapple-like, brownish, wrinkled and musky. Contains 1 rough seed. Fruit eaten raw and cooked or preserved in various ways. Grown from seed in Florida. Crown-budded in India where the fruit is prized.

Other Publications
on Florida Plants

(Those out of print may be consulted in public libraries)

Ames, B. and O. *Drawings of Florida Orchids,* Botanical Museum, Harvard University, Cambridge, Mass. 1947.

Baker, M. F. *Florida Wild Flowers.* Macmillan Co., N. Y. 1949. Reprinted by Horticultural Books, Inc., Stuart, Fla. 1972.

Birdsey, M. R. *Cultivated Aroids,* Gillick Press, Berkeley, Calif. 1951.

Brown, B. F. *Florida's Beautiful Crotons.* Published by author, 244 Michigan Ave., Indialantic, Fla. 1960.

Bush, C. S. *Flowers, Shrubs and Trees for Florida Homes.* Bul. 195. Florida State Dept. Agric., Tallahassee, Fla. 1969.

Bush, C. S. and J. F. Morton. *Native Trees and Plants for Florida Landscaping.* Bul. 193. Florida State Dept. Agric., Tallahassee, Fla. 1968. 2d ed., 1972. Revised 1978.

Buswell, W. M. *Native Orchids of South Florida.* Univ. of Miami Press, Coral Gables, Fla. 1945 (out of print).

Buswell, W. M. *Native Shrubs of South Florida.* Univ. of Miami Press, Coral Gables, Fla. 1946.

Buswell, W. M. *Native Trees and Palms of South Florida.* Univ. of Miami Press, Coral Gables, Fla. 1945. (out of print)

Conover, C. A. and E. W. McElwee. *Selected Trees for Florida Homes.* Bul. 182, Univ. of Florida, Inst. of Food & Agric. Sci., Gainesville, Fla. 1971.

Craighead, F. C. *Orchids and Other Air Plants of the Everglades National Park.* Univ. of Miami Press, Coral Gables, Fla. 1963.

Dickey, R. D. and H. Mowry. *Ornamental Hedges for Florida.* Bul. 178B. Univ. of Florida, Inst. of Food & Agric. Sci., Gainesville, Fla. 1969.

Ledin, R. B. *Compositae of South Florida.* Proc. Fla. Acad. Sci. 1951. Reprinted by Univ. of Miami Press, Coral Gables, Fla. (out of print)

Long, R. W. and O. Lakela, *A Flora of Tropical Florida.* Univ. of Miami Press, Coral Gables, Fla. 1971.

Morton, J. F. *Exotic Plants.* Golden Press, New York. 1971.

Morton, J. F. *Plants Poisonous to People in Florida and Other Warm Areas.* Fairchild Tropical Garden, Miami, Fla. 1977.

Morton, J. F. *Plants Poisonous to People: Wall Charts I & II.* Trend House, Tampa, Fla. 1968.

Morton, J. F. *Some Useful and Ornamental Plants of the Caribbean Gardens.* Caribbean Gardens, Naples, Fla. 1955.

Morton, J. F. *Wild Plants for Survival in South Florida.* Fairchild Tropical Garden, Miami, Fla. 4th ed. 1977.

Morton, K. P. and J. F. *Fifty Tropical Fruits of Nassau* (All grow in Florida), Text House, Inc., Coral Gables, Fla. 1946 (out of print)

Nehrling, H. *My Garden In Florida,* Vols. 1 & 2. American Eagle, Estero, Fla. 1944, 1946 (out of print)

Palmer, K. and M. *Hibiscus Unlimited and How to Know Them.* Creative Press, Inc., St. Petersburg, Fla. Rev'd.

Perry, Mac. *Landscape Your Florida Home.* E. A. Seemann Publishing, Inc., Miami, Fla. 1972.

Simpson, C. T. *Ornamental Gardening in Florida.* Pub'd by the author, Little River, Fla. Rev'd ed. 1926 (out of print).

Smiley, N. *Florida Gardening Month by Month.* Univ. of Miami Press, Coral Gables, Fla. Revised ed., 1970.

Smiley, N. *Tropical Planting and Gardening for South Florida and the West Indies.* Univ. of Miami Press, Coral Gables, Fla. 1961. 4th ptg. 1970.

Sturrock, D. *Fruits for Southern Florida.* Southeastern Printing Co., Inc., Stuart, Fla. 1959.

Watkins, J. V. *Florida Landscape Plants, Native and Exotic.* Univ. of Florida Press, Gainesville, Fla. 1969.

Watkins, J. V. and H. S. Wolfe. *Your Florida Garden.* Univ. of Florida Press, Gainesville, Fla. 5th ed. 1968.

West, E. and L. Arnold. *Native Trees of Florida.* Univ. of Florida Press, Gainesville, Fla. Rev'd 1956.

INDEX

of common names and botanical synonyms
(The preferred botanical names will be found in alphabetical sequence in the text)